P9-DWZ-954

SCIENCE AND HUMAN VALUES

SCIENCE AND HUMAN VALUES

By the same author

SCIENCE AND HUMAN VALUES

Revised Edition
with a new dialogue
THE ABACUS AND THE ROSE

by J. Bronowski

PERENNIAL LIBRARY

Harper & Row, Publishers, New York
Grand Rapids, Philadelphia, St. Louis, San Francisco
London, Singapore, Sydney, Tokyo, Toronto

L.C° SOUTH CAMPUS LIBRARY

Q
171
.B8785
1990

Reprinted by arrangement with Julian Messner, Inc.

SCIENCE AND HUMAN VALUES. Copyright © 1956 by J. Bronowski. Revised edition copyright © 1965 by J. Bronowski. All rights reserved. Printed in the United States of America. No. part of this book may be used or reproduced in any manner whatsoever without written permission except in the case of brief quotations embodied in critical articles and reviews. For information address Harper & Row, Publishers, Inc., 10 East 53rd Street, New York, N.Y. 10022..

First PERENNIAL LIBRARY edition published 1972.

Library of Congress Cataloging-in-Publication Data

Bronowski, Jacob, 1908–1974.
    Science and human values / by J. Bronowski. — Rev. ed. with a new dialogue, The abacus and the rose.
    p. cm.
    ISBN 0-06-097281-5
    1. Science.   I. Title.   II. Title: Abacus and the rose.
    Q171.B8785 1990   89-45631
    500—dc20

   00 **RRD H** 30

MAR 10 2004

Contents

Contents

Illustrations

Preface to the Revised Edition

THE three essays which make up *Science and Human Values* were first given as lectures at the Massachusetts Institute of Technology on 26 February, 5 March and 19 March 1953, when I was Carnegie Professor there. They were published after my return to England as articles in three issues of the *Universities Quarterly* in 1956, and a little later *The Nation* in America gave up its last issue of that year entirely to them. At that time the essays still seemed to me to concern only scientists, but general interest in them continued to grow, and passages and ideas from them became current both in England and in America. They were first published as a book in America in 1958 and in England in 1961.

In the years that have passed since I first put forward the ideas in these essays I have continued to practice science, and to think about its relation to the values by which we live. Yet, reading these essays again now, I find little that I would change and few things that I want to add. I have therefore made only small changes in the text, and I have put the added material mostly into notes at the end of each essay. These notes fill out some points which have given rise to discussion, and they also quote relevant new matter which has appeared since the essays were first printed. The notes amplify my text, and underline it; and I do not want to change anything else.

At the end of *Science and Human Values,* I have also added a new dialogue, *The Abacus and the Rose,* which is essentially an extended note. It discusses the theme, which runs througout the essays, that science is as integral a part of the culture of our age as the arts are. This theme was summed up by Sir Charles Snow in his Rede Lecture in 1959 in the telling phrase *The Two Cultures*. Since then it has been debated with so much passion that it has seemed to me natural and just to present the arguments in the dramatic form of a dialogue.

It happens that there exists a classical model for such a dialogue. This is the *Dialogue on the Great World Systems* which Galileo published in 1632 and which he was forced to recant in the following year. True, the two world systems which confront one another in Galileo's dialogue are, in appearance, merely different scientific arrangements: the difference between them is a technical nicety, whether the earth is the center of the universe or not. But of course, the deep issue between the systems was not technical, and it divided the culture of the age as harshly as ours is still divided. The trial of Galileo in 1633 was a spectacular display of strength by the forces of tradition in the Holy Office, and they kept his *Dialogue* on the Index of forbidden books for another two hundred years. This was the most tenacious rearguard action that has yet been fought by established belief against the challenging spirit of science. But it was not the last such bitter action, and the form of the dialogue in *The Abacus and the Rose* implies that.

I have tried to make the dialogue do its work, and to put the arguments on each side fairly and with pleasure, in words which do not caricature its case. But where I have doubted my ability, I have thought it better to quote a criticism in the robust phrases which the critic himself has used. I have drawn in particular on the Richmond Lecture which Dr. F. R. Leavis gave in 1962.

The Abacus and the Rose: A New Dialogue on Two World Systems was broadcast in the Third Programme of the British Broadcasting Corporation on 6 November 1962 as one of the programs to celebrate the fortieth anniversary of the Corporation. It was first published by *The Nation* as its New Year issue for 1964. Its publication now with *Science and Human Values* gives me the happy occasion to thank *The Nation* and its editor, Carey McWilliams, for their sustained interest in the heart-searchings of a scientist.

In only one respect would I want to enlarge what I have said in this book about science and human values, if I were starting afresh today to write about their relation for the first time. In the essays as I have written them I have deliberately confined myself to establishing one central proposition: that the practice of science compels the practitioner to form for himself a fundamental set of universal values. I have not suggested that this set embraces all the human values; I was sure when I wrote that it did not; but at the time I did not want to blur the argument by discussing the whole spectrum of values. Now that the crux of my argument has been accepted, I would, were I beginning again, give some space also to a discussion of those values which are not generated by the practice of science—the values of tenderness, of kindliness, of human intimacy and love. These form a different domain from the sharp and, as it were, Old Testament virtues which science produces, but of course they do not negate the values of science. I shall hope to write about the relation between the two sets of values at another time, and to show how we need to link them in our behavior.

As it is, I leave *Science and Human Values* at the point which its last page reaches: the demonstration that values are not rules,

> *but are those deeper illuminations in whose light justice and unjustice, good and evil, means and ends are seen in fearful sharpness of outline.*

This thought, that the exactness of science can give a context for our judgments, seems to me as important in its application today as it was when first I wrote it.

The gravest indictment that can be made of our general
ized culture is, in fact, that it erodes our sense of the contex
in which judgments must be made. Let me end with a practi
cal example. When I returned from the physical shock o
Nagasaki, which I have described in the first page of thi
book, I tried to persuade my colleagues in governments and
in the United Nations that Nagasaki should be preserved ex
actly as it was then. I wanted all future conferences on dis
armament, and on other issues which weigh the fates o
nations, to be held in that ashy, clinical sea of rubble. I stil
think as I did then, that only in this forbidding context could
statesmen make realistic judgments of the problems which
they handle on our behalf. Alas, my official colleague
thought nothing of my scheme; on the contrary, they pointed
out to me that delegates would be uncomfortable in Naga
saki.

J.B

The Salk Institute for Biological Studies
San Diego, California
February, 1964

SCIENCE AND HUMAN VALUES

SCIENCE AND HUMAN VALUES

one

The Creative Mind

ON a fine November day in 1945, late in the afternoon, I was landed on an airstrip in southern Japan. From there a jeep was to take me over the mountains to join a ship which lay in Nagasaki Harbor. I knew nothing of the country or the distance before us. We drove off; dusk fell; the road rose and fell away, the pine woods came down to the road, straggled on and opened again. I did not know that we had left the open country until unexpectedly I heard the ship's loudspeakers broadcasting dance music. Then suddenly I was aware that we were already at the center of damage in Nagasaki. The shadows behind me were the skeletons of the Mitsubishi factory buildings, pushed backwards and sideways as if by a giant hand. What I had thought to be broken rocks was a concrete power house with its roof punched in. I could now make out the outline of two crumpled gasometers; there was a cold furnace festooned with service pipes; otherwise nothing but cockeyed telegraph poles and loops of wire in a bare waste of ashes. I had blundered into this desolate lanscape as instantly as one might wake among the craters of the moon. The moment of recognition when I realized that I was already in Nagasaki is present to me as I write, as vividly as when I lived it. I see the warm night and the meaningless shapes; I can even remember the tune that was coming from the ship. It was a dance tune which had been popular in 1945, and it was called 'Is You Is Or Is You Ain't Ma Baby?'

These essays, which I have called *Science and Human Values*, were born at that moment. For the moment I have recalled was a universal moment; what I met was, almost as abruptly, the experience of mankind. On an evening like that evening, some time in 1945, each of us in his own way learned that his imagination had been dwarfed. We looked up and saw the power of which we had been proud loom over us like the ruins of Nagasaki.

The power of science for good and for evil has troubled other minds than ours. We are not here fumbling with a new dilemma; our subject and our fears are as old as the tool-making civilizations. Men have been killed with weapons before now: what happened at Nagasaki was only more massive (for 40,000 were killed there by a flash which lasted seconds) and more ironical (for the bomb exploded over the main Christian community in Japan). Nothing happened in 1945 except that we changed the scale of our indifference to man; and conscience, in revenge, for an instant became immediate to us. Before this immediacy fades in a sequence of televised atomic tests, let us acknowledge our subject for what it is: civilization face to face with its own implications. The implications are both the industrial slum which Nagasaki was before it was bombed, and the ashy desolation which the bomb made of the slum. And civilization asks of both ruins, 'Is You Is Or Is You Ain't Ma Baby?'

2

The man whom I imagine to be asking this question, wrily with a sense of shame, is not a scientist; he is civilized man. It is of course more usual for each member of civilization to take flight from its consequences by protesting that others have failed him. Those whose education and perhaps tastes have confined them to the humanities protest that the scientists alone are to blame, for plainly no mandarin ever made a bomb or an industry. The scientists say, with equal contempt, that the Greek scholars and the earnest cataloguers of cave paintings do well to wash their hands of blame; but what in fact are they doing to help direct the society whose ills grow more often from inaction than from error?

This absurd division reached its *reductio ad absurdum*, I think, when one of my teachers, G. H. Hardy, justified his great life work on the ground that it could do no one the least harm—or the least good.[1] But Hardy was a mathematician; will humanists really let him opt out of the conspiracy of scientists? Or are scientists in their turn to forgive Hardy because, protest as he might, most of them learned their indispensable mathematics from his books?

There is no comfort in such bickering. When Shelley pictured science as a modern Prometheus who would wake the world to a wonderful dream of Godwin, he was alas too simple. But it is as pointless to read what has happened since as a nightmare. Dream or nightmare, we have to live our experience as it is, and we have to live it awake. We live in a world which is penetrated through and through by science, and which is both whole and real. We cannot turn it into a game simply by taking sides.

And this make-believe game might cost us what we value most: the human content of our lives. The scholar who disdains science may speak in fun, but his fun is not quite a laughing matter. To think of science as a set of special tricks, to see the scientist as the manipulator of outlandish skills— this is the root of the poison mandrake which flourishes rank in the comic strips. There is no more threatening and no more degrading doctrine than the fancy that somehow we may shelve the responsibility for making the decisions of our society by passing it to a few scientists armored with a special magic. This is another dream, the dream of H. G. Wells, in which the tall elegant engineers rule, with perfect benevolence, a humanity which has no business except to be happy. To H. G. Wells, this was a dream of heaven — a modern version of the idle, harp-resounding heaven of other childhood pieties. But in fact it is the picture of a slave society, and should make us shiver whenever we hear a man of sensibility dismiss science as someone else's concern. The world today is made, it is powered by science; and for any man to abdicate an interest in science is to walk with open eyes towards slavery.

My aim in this book is to show that the parts of civilization make a whole: to display the links which give society its coherence and, more, which give it life. In particular, I want to show the place of science in the canons of conduct which it has still to perfect.

This subject falls into three parts. The first is a study of the nature of the scientific activity, and with it of all those imaginative acts of understanding which exercise 'The Creative Mind.' After this it is logical to ask what is the nature of the truth, as we seek it in science and in social life; and to trace the influence which this search for empirical truth has had on conduct. This influence has prompted me to call the second part 'The Habit of Truth.' Last I shall study the conditions for the success of science, and find in them the values of man which science would have had to invent afresh if man had not otherwise known them: the values which make up 'The Sense of Human Dignity.'

This, then, is a high-ranging subject which is not to be held in the narrow limits of a laboratory. It disputes the prejudice of the humanist who takes his science sourly and, equally, the petty view which many scientists take of their own activity and that of others. When men misunderstand their own work, they cannot understand the work of others; so that it is natural that these scientists have been indifferent to the arts. They have been content, with the humanists, to think science mechanical and neutral; they could therefore justify themselves only by the claim that it is practical. By this lame criterion they have of course found poetry and music and painting at least unreal and often meaningless. I challenge all these judgments.

3

There is a likeness between the creative acts of the mind in art and in science. Yet, when a man uses the word science in such a sentence, it may be suspected that he does not mean what the headlines mean by science. Am I about to idle away to those riddles in the Theory of Numbers which Hardy loved, or to the heady speculations of astrophysicists, in order to make claims for abstract science which have no bearing on its daily practice?

I have no such design. My purpose is to talk about science as it is, practical and theoretical. I define science as the organization of our knowledge in such a way that it commands more of the hidden potential in nature. What I have in mind therefore is both deep and matter of fact; it reaches from the kinetic theory of gases to the telephone and the suspension bridge and medicated toothpaste. It admits no sharp boundary between knowledge and use. There are of course people who like to draw a line between pure and applied science; and oddly, they are often the same people who find art unreal. To them, the word useful is a final arbiter, either for or against a work; and they use this word as if it can mean only what makes a man feel heavier after meals.

There is no sanction for confining the practice of science in this or another way. True, science is full of useful inventions. And its theories have often been made by men whose imagination was directed by the uses to which their age looked. Newton turned naturally to astronomy because it was the subject of his day, and it was so because finding one's way at sea had long been a practical preoccupation of the society into which he was born. It should be added, mischievously, that astronomy also had some standing because it was used very practically to cast horoscopes. (Kepler used it for this purpose; in the Thirty Years' War he cast the horo-

scope of Wallenstein which wonderfully told his character,
and he predicted a universal disaster for 1634 which proved
to be the murder of Wallenstein.[2])

In a setting which is more familiar, Faraday worked all
his life to link electricity with magnetism because this was
the glittering problem of his day; and it was so because his
society, like ours, was on the lookout for new sources of
power. Consider a more modest example today: the new
mathematical methods of automatic control, a subject sometimes called cybernetics, have been developed now because
this is a time when communication and control have in effect
become forms of power.[3] These inventions have been directed by social needs, and they are useful inventions; yet it
was not their usefulness which dominated and set light to
the minds of those who made them. Neither Newton nor
Faraday, nor yet Norbert Wiener, spent their time in a
scramble for patents.

What a scientist does is compounded of two interests:
the interest of his time and his own interest. In this his
behavior is no different from any other man's. The need
of the age gives its shape to scientific progress as a whole.
But it is not the need of the age which gives the individual
scientist his sense of pleasure and of adventure, and that
excitement which keeps him working late into the night
when all the useful typists have gone home at five o'clock.
He is personally involved in his work, as the poet is in his,
and as the artist is in the painting. Paints and painting too
must have been made for useful ends; and language was
developed, from whatever beginnings, for practical communication. Yet you cannot have a man handle paints or language or the symbolic concepts of physics, you cannot even
have him stain a microscope slide, without instantly waking
in him a pleasure in the very language, a sense of exploring
his own activity. This sense lies at the heart of creation.[4]

4

The sense of personal exploration is as urgent, and as delightful, to the practical scientist as to the theoretical. Those who think otherwise are confusing what is practical with what is humdrum. Good humdrum work without originality is done every day by everyone, theoretical scientists as well as practical, and writers and painters too, as well as truck drivers and bank clerks. Of course the unoriginal work keeps the world going; but it is not therefore the monopoly of practical men. And neither need the practical man be unoriginal. If he is to break out of what has been done before, he must bring to his own tools the same sense of pride and discovery which the poet brings to words. He cannot afford to be less radical in conceiving and less creative in designing a new turbine than a new world system.

And this is why in turn practical discoveries are not made only by practical men. As the world's interest has shifted, since the Industrial Revolution, to the tapping of new springs of power, the theoretical scientist has shifted his interests too. His speculations about energy have been as abstract as once they were about astronomy; and they have been profound now as they were then, because the man loved to think. The Carnot cycle and the dynamo grew equally from this love, and so did nuclear physics and the German V weapons and Kelvin's interest in low temperatures. Man does not invent by following either use or tradition; he does not invent even a new form of communication by calling a conference of communication engineers. Who invented the television set? In any deep sense, it was Clerk Maxwell who foresaw the existence of radio waves, and Heinrich Hertz who proved it, and J. J. Thomson who discovered the electron. This is not said in order to rob any practical man of the invention, but from a sad sense of justice; for neither Maxwell nor Hertz nor J. J. Thomson would take pride in television just now.

Man masters nature not by force but by understanding. This is why science has succeeded where magic failed: because it has looked for no spell to cast over nature. The alchemist and the magician in the Middle Ages thought, and the addict of comic strips is still encouraged to think, that nature must be mastered by a device which outrages her laws. But in four hundred years since the Scientific Revolution we have learned that we gain our ends only *with* the laws of nature; we control her only by understanding her laws. We cannot even bully nature by any insistence that our work shall be designed to give power over her. We must be content that power is the byproduct of understanding. So the Greeks said that Orpheus played the lyre with such sympathy that wild beasts were tamed by the hand on the strings. They did not suggest that he got this gift by setting out to be a lion tamer.

5

What is the insight with which the scientist tries to see into nature? Can it indeed be called either imaginative or creative? To the literary man the question may seem merely silly. He has been taught that science is a large collection of facts; and if this is true, then the only seeing which scientists need do is, he supposes, seeing the facts. He pictures them, the colorless professionals of science, going off to work in the morning into the universe in a neutral, unexposed state. They then expose themselves like a photographic plate. And then in the darkroom or laboratory they develop the image, so that suddenly and startlingly it appears, printed in capital letters, as a new formula for atomic energy.

Men who have read Balzac and Zola are not deceived by the claims of these writers that they do no more than

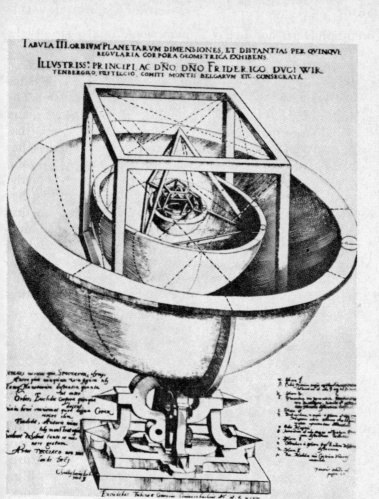

The Planetary Orbits Embedded in the Regular Solids
from Kepler's Mysterium Cosmographicum, 1596

record the facts. The readers of Christopher Isherwood do not take him literally when he writes 'I am a camera.' Yet the same readers solemnly carry with them from their school-days this foolish picture of the scientist fixing by some mechanical process the facts of nature. I have had of all people a historian tell me that science is a collection of facts, and his voice had not even the ironic rasp of one filing cabinet reproving another.

It seems impossible that this historian had ever studied the beginnings of a scientific discovery. The Scientific Revolution can be held to begin in the year 1543 when there was brought to Copernicus, perhaps on his deathbed, the first printed copy of the book he had finished about a dozen years earlier. The thesis of this book is that the earth moves around the sun. When did Copernicus go out and record this fact with his camera? What appearance in nature prompted his outrageous guess? And in what odd sense is this guess to be called a neutral record of fact?

Less than a hundred years after Copernicus, Kepler published (between 1609 and 1619) the three laws which describe the paths of the planets. The work of Newton and with it most of our mechanics spring from these laws. They have a solid, matter of fact sound. For example, Kepler says that if one squares the year of a planet, one gets a number which is proportional to the cube of its average distance from the sun. Does anyone think that such a law is found by taking enough readings and then squaring and cubing everything in sight? If he does then, as a scientist, he is doomed to a wasted life; he has as little prospect of making a scientific discovery as an electronic brain has.

It was not this way that Copernicus and Kepler thought, or that scientists think today. Copernicus found that the orbits of the planets would look simpler if they were looked at from the sun and not from the earth. But he did not in the first place find this by routine calculation. His first step was a leap of imagination—to lift himself from the earth, and put himself wildly, speculatively into the sun.[5] 'The earth conceives from the sun,' he wrote; and 'the sun rules the family of stars.' We catch in his mind an image, the gesture of the virile man standing in the sun, with arms outstretched, overlooking the planets. Perhaps Copernicus took the picture from the drawings of the youth with outstretched arms which the Renaissance teachers put into their books on the proportions of the body. Perhaps he had seen Leonardo's drawings of his loved pupil Salai. I do not know. To me, the gesture of Copernicus, the shining youth looking outward from the sun, is still vivid in a drawing which William Blake in 1780 based on all these: the drawing which is usually called *Glad Day*.[6]

Kepler's mind, we know, was filled with just such fanciful analogies; and we know what they were. Kepler wanted to relate the speeds of the planets to the musical intervals. He tried to fit the five regular solids into their orbits. None of these likenesses worked, and they have been forgotten; yet they have been and they remain the stepping stones of every creative mind. Kepler felt for his laws by way of metaphors, he searched mystically for likenesses with what he knew in every strange corner of nature. And when among these guesses he hit upon his laws, he did not think of their numbers as the balancing of a cosmic bank account, but as a revelation of the unity in all nature. To us, the analogies by which Kepler listened for the movement of the planets in the music of the spheres are farfetched.[7] Yet are they more so than the wild leap by which Rutherford and Bohr in our own century found a model for the atom in, of all places, the planetary system?

6

No scientific theory is a collection of facts. It will not even do to call a theory true or false in the simple sense in which every fact is either so or not so. The Epicureans held that matter is made of atoms two thousand years ago and we are now tempted to say that their theory was true. But if we do so we confuse their notion of matter with our own. John Dalton in 1808 first saw the structure of matter as we do today, and what he took from the ancients was not their theory but something richer, their image: the atom. Much of what was in Dalton's mind was as vague as the Greek notion, and quite as mistaken. But he suddenly gave life to the new facts of chemistry and the ancient theory together, by fusing them to give what neither had: a coherent picture of how matter is linked and built up from different kinds of atoms. The act of fusion is the creative act.

All science is the search for unity in hidden likenesses. The search may be on a grand scale, as in the modern theories which try to link the fields of gravitation and electromagnetism. But we do not need to be browbeaten by the scale of science. There are discoveries to be made by snatching a small likeness from the air too, if it is bold enough. In 1935 the Japanese physicist Hideki Yukawa wrote a paper which can still give heart to a young scientist. He took as his starting point the known fact that waves of light can sometimes behave as if they were separate pellets. From this he reasoned that the forces which hold the nucleus of an atom together might sometimes also be observed as if they were solid pellets. A schoolboy can see how thin Yukawa's analogy is, and his teacher would be severe with it. Yet Yukawa without a blush calculated the mass of the pellet he expected to see, and waited. He was right; his meson was found, and a range of other mesons, neither the existence nor the nature of which had been suspected before. The likeness had borne fruit.

The scientist looks for order in the appearances of nature by exploring such likenesses. For order does not display itself of itself; if it can be said to be there at all, it is not there for the mere looking. There is no way of pointing a finger or a camera at it; order must be discovered and, in a deep sense, it must be created. What we see, as we see it, is mere disorder.

This point has been put trenchantly in a fable by Karl Popper. Suppose that someone wished to give his whole life to science. Suppose that he therefore sat down, pencil in hand, and for the next twenty, thirty, forty years recorded in notebook after notebook everything that he could observe. He may be supposed to leave out nothing: today's humidity, the racing results, the level of cosmic radiation and the stockmarket prices and the look of Mars, all would be there. He would have compiled the most careful record of nature that has ever been made; and, dying in the calm certainty of a life well spent, he would of course leave his notebooks to the Royal Society. Would the Royal Society thank him for the treasure of a lifetime of observation? It would not. The Royal Society would treat his notebooks exactly as the English bishops have treated Joanna Southcott's box. It would refuse to open them at all, because it would know without looking that the notebooks contain only a jumble of disorderly and meaningless items.

7

Science finds order and meaning in our experience, and sets about this in quite a different way. It sets about it as Newton did in the story which he himself told in his old age, and of which the schoolbooks give only a caricature. In the year 1665, when Newton was twenty-two, the plague broke out in southern England, and the University of Cambridge was closed. Newton therefore spent the next eighteen months at home, removed from traditional learning, at a time when he was impatient for knowledge and, in his own phrase, 'I was in the prime of my age for invention.' In this eager, boy-ish mood, sitting one day in the garden of his widowed mother, he saw an apple fall. So far the books have the story right; we think we even know the kind of apple; tradition has it that it was a Flower of Kent. But now they miss the crux of the story. For what struck the young New-ton at the sight was not the thought that the apple must be drawn to the earth by gravity; that conception was older than Newton. What struck him was the conjecture that the same force of gravity, which reaches to the top of the tree, might go on reaching out beyond the earth and its air, end-lessly into space. Gravity might reach the moon: this was Newton's new thought; and it might be gravity which holds the moon in her orbit. There and then he calculated what force from the earth (falling off as the square of the distance) would hold the moon, and compared it with the known force of gravity at tree height. The forces agreed; Newton says laconically, 'I found them answer pretty nearly.' Yet they agreed only nearly: the likeness and the approximation go together, for no likeness is exact. In Newton's sentence modern science is full grown.

It grows from a comparison. It has seized a likeness be-tween two unlike appearances; for the apple in the summer garden and the grave moon overhead are surely as unlike in their movements as two things can be. Newton traced in them two expressions of a single concept, gravitation: and the concept (and the unity) are in that sense his free creation. The progress of science is the discovery at each step of a new order which gives unity to what had long seemed unlike. Faraday did this when he closed the link between electricity and magnetism. Clerk Maxwell did it when he linked both with light. Einstein linked time with space, mass with energy, and the path of light past the sun with the flight of a bullet; and spent his dying years in trying to add to these likenesses another, which would find a single imaginative order between the equations of Clerk Maxwell and his own geometry of gravitation.

8

When Coleridge tried to define beauty, he returned always to one deep thought: beauty, he said, is 'unity in variety.' [8] Science is nothing else than the search to discover unity in the wild variety of nature—or more exactly, in the variety of our experience. Poetry, painting, the arts are the same search, in Coleridge's phrase, for unity in variety. Each in its own way looks for likenesses under the variety of human experience. What is a poetic image but the seizing and the exploration of a hidden likeness, in holding together two parts of a comparison which are to give depth each to the other? When Romeo finds Juliet in the tomb, and thinks her dead, he uses in his heartbreaking speech the words,

> Death that hath suckt the honey of thy breath.

The critic can only haltingly take to pieces the single shock which this image carries. The young Shakespeare admired Marlowe, and Marlowe's Faustus had said of the ghostly kiss of Helen of Troy that it sucked forth his soul. But that is a pale image; what Shakespeare has done is to fire it with the single word honey. Death is a bee at the lips of Juliet, and the bee is an insect that stings; the sting of death was a commonplace phrase when Shakespeare wrote. The sting is there, under the image; Shakespeare has packed it into the word honey; but the very word rides powerfully over its own undertones. Death is a bee that stings other people, but it comes to Juliet as if she were a flower; this is the moving thought under the instant image. The creative mind speaks in such thoughts.

The poetic image here is also, and accidentally, heightened

by the tenderness which town dwellers now feel for country
ways. But it need not be; there are likenesses to conjure with,
and images as powerful, within the man-made world. The
poems of Alexander Pope belong to this world. They are not
countrified, and therefore readers today find them unemo-
tional and often artificial. Let me then quote Pope: here he is
in a formal satire face to face, towards the end of his life, with
his own gifts. In eight lines he looks poignantly forward towards death and back to the laborious years which made him
famous.

> Years foll'wing Years, steal something ev'ry day,
> At last they steal us from our selves away;
> In one our Frolicks, one Amusements end,
> In one a Mistress drops, in one a Friend:
> This subtle Thief of Life, this paltry Time,
> What will it leave me, if it snatch my Rhime?
> If ev'ry Wheel of that unweary'd Mill
> That turn'd ten thousand Verses, now stands still.

The human mind had been compared to what the eighteenth
century called a mill, that is to a machine, before; Pope's own
idol Bolingbroke had compared it to a clockwork. In these
lines the likeness goes deeper, for Pope is thinking of the
ten thousand Verses which he had translated from Homer:
what he says is sad and just at the same time, because this
really had been a mechanical and at times a grinding task.[9]
Yet the clockwork is present in the image too; when the
wheels stand still, time for Pope will stand still for ever; we
feel that we already hear, over the horizon, Faust's defiant
reply to Mephistopheles, which Goethe had not yet written
—'let the clock strike and stop, let the hand fall, and time
be at an end.'

> Werd ich zum Augenblicke sagen:
> Verweile doch! du bist so schön!

Dann magst du mich in Fesseln schlagen,
Dann will ich gern zugrunde gehn!
Dann mag die Totenglocke schallen,
Dann bist du deines Dienstes frei,
Die Uhr mag stehn, der Zeiger fallen,
Es sei die Zeit für mich vorbei! [10]

I have quoted Pope and Goethe because their metaphor
here is not poetic; it is rather a hand reaching straight into
experience and arranging it with new meaning. Metaphors
of this kind need not always be written in words. The most
powerful of them all is simply the presence of King Lear
and his Fool in the hovel of a man who is shamming madness,
while lightning rages outside. Or let me quote another clash
of two conceptions of life, from a modern poet. In his later
poems W. B. Yeats was troubled by the feeling that in shut-
ting himself up to write, he was missing the active pleasures
of life; and yet it seemed to him certain that the man who
lives for these pleasures will leave no lasting work behind
him. He said this at times very simply, too:

The intellect of man is forced to choose
Perfection of the life, or of the work.

This problem, whether a man fulfills himself in work or
in play, is of course more common than Yeats allowed; and
it may be more commonplace. But it is given breadth and
force by the images in which Yeats pondered it.

Get all the gold and silver that you can,
Satisfy ambition, or animate
The trivial days and ram them with the sun,
And yet upon these maxims meditate:
All women dote upon an idle man
Although their children need a rich estate;
No man has ever lived that had enough
Of children's gratitude or woman's love.[11]

The love of women, the gratitude of children: the images fix two philosophies as nothing else can. They are tools of creative thought, as coherent and as exact as the conceptual images with which science works: as time and space, or as the proton and the neutron.

9

The discoveries of science, the works of art are explorations — more, are explosions, of a hidden likeness. The discoverer or the artist presents in them two aspects of nature and fuses them into one. This is the act of creation, in which an original thought is born, and it is the same act in original science and original art. But it is not therefore the monopoly of the man who wrote the poem or who made the discovery. On the contrary, I believe this view of the creative act to be right because it alone gives a meaning to the act of appreciation. The poem or the discovery exists in two moments of vision: the moment of appreciation as much as that of creation; for the appreciator must see the movement, wake to the echo which was started in the creation of the work. In the moment of appreciation we live again the moment when the creator saw and held the hidden likeness. When a simile takes us aback and persuades us together, when we find a juxtaposition in a picture both odd and intriguing, when a theory is at once fresh and convincing, we do not merely nod over someone else's work. We re-enact the creative act, and we ourselves make the discovery again. At bottom, there is no unifying likeness there until we too have seized it, we too have made it for ourselves.

How slipshod by comparison is the notion that either art or science sets out to copy nature. If the task of the painter were to copy for men what they see, the critic could make only a single judgment: either that the copy is right or that it is wrong. And if science were a copy of fact, then every theory would be either right or wrong, and would be so for ever. There would be nothing left for us to say but this is so, or is not so. No one who has read a page by a good critic or a speculative scientist can ever again think that this barren choice of yes or no is all that the mind offers.

Reality is not an exhibit for man's inspection, labelled 'Do not touch.' There are no appearances to be photographed, no experiences to be copied, in which we do not take part. Science, like art, is not a copy of nature but a re-creation of her. We re-make nature by the act of discovery, in the poem or in the theorem. And the great poem and the deep theorem are new to every reader, and yet are his own experiences, because he himself re-creates them. They are the marks of unity in variety; and in the instant when the mind seizes this for itself, in art or in science, the heart misses a beat.

NOTES

1. This is the argument in Hardy's delightful small book *A Mathematician's Apology* (Cambridge, 1940). The title of the book and its date, soon after the outbreak of war, suggest that it was prompted by the same distress at the visible misuse of science that has prompted my book. Hardy was a great mathematician, and he was also (under his restrained manner) a man of passionate human and social sympathies.

2. Wallenstein, the great Catholic general in this religious war, was born at the conjunction of Jupiter and Saturn. There is a conjunction of these two planets when they appear to change places repeatedly: in the language of astrology, when they play. This rare conjunction has foreshadowed a number of historic events, for historic events are always plentiful. It occurred about six years before the birth of Christ, and also before the Protestant king William of Orange came to England in the Glorious Revolution of 1688. The same conjunction of Jupiter and Saturn occurred in 1940, and I will quote a poem which I wrote about it in that year.

Jupiter and Saturn played.
The age was broken and re-made.
A rocket rose from Bethlehem.
Christ marched with the Orangemen;
Till, diving, the exploding light
Struck today, and charred it white.

The rocket roars and plunges out.
Saturn and Jupiter turn about.
No child again shall put to shame
The gunsights trained on Bethlehem;
While ice-cap, omen, march to birth
The orbit of the screaming earth.

It was in observing another rare astrological event, the triple conjunction of Jupiter, Saturn and Mars, that Joannes Brunowski discovered, and informed Kepler of, the famous supernova of 1604.

3. At the time of the Scientific Revolution in the sixteenth century, and for two centuries after it, most self-made men got their wealth by trade (in which I include the support of trade by insurance and banking), and often by oversea trade. As *The Merchant of Venice* reminds us, this is how the great fortunes in North Italy, in Holland and in England were made. It was therefore natural that science in these two centuries was agog with problems of trade, and particularly of navigation. The Industrial Revolution in the eighteenth century shifted the source of wealth from trade to manufacture; and manufacture has needed more and more mechanical energy. Science has therefore been preoccupied in the last two centuries with problems which center on energy—practical problems from the heat engine to the electromagnetic field, and theoretical problems from thermodynamics to atomic structure. Now that we are in sight of having as much energy as we can need, the interest of scientists is moving from the generation of energy to its control, and particularly to the automatic control of power processes, whose tools are the valve, the semi-conductor and the computer. A characteristic invention of the Scientific Revolution was the telescope, of which Galileo heard from Holland, and which he presented to the Doge after a demonstration in the port of Venice in the presence of the Senate in 1609. The characteristic invention of the Industrial Revolution was the power machine which does the routine work of the human muscle. The characteristic invention of the second Industrial Revolution through which we are passing is the control mechanism which does the routine work of the human brain.

4. As an example, consider the practice of mathematics. Mathematics is in the first place a language in which we discuss those parts of the real world which can be described by numbers or by similar relations of order. But with the workaday business of translating the facts into this language there naturally goes, in those who are good at it, a pleasure in the activity itself.

They find the language richer than its bare content; what is translated comes to mean less to them than the logic and the style of saying it; and from these overtones grows mathematics as a literature in its own right. Mathematics in this sense, pure mathematics, is a form of poetry, which has the same relation to the prose of practical mathematics as poetry has to prose in any other language. This element of poetry, the delight in exploring the medium for its own sake, is an essential ingredient in the creative process.

5. This has now been admirably documented by Thomas S. Kuhn in *The Copernican Revolution* (Harvard, 1957). As he shows, from the Neoplatonist elements in the new humanism 'some Renaissance scientists, like Copernicus, Galileo, and Kepler, seem to have drawn two decidedly un-Aristotelian ideas: a new belief in the possibility and importance of discovering simple arithmetic and geometric regularities in nature, and a new view of the sun as the source of all vital principles and forces in the universe.' Kuhn draws particular attention to the influence of the 'symbolic identification of the sun and God' in the *Liber de Sole* of Marsilio Ficino, a central figure (with Pico della Mirandola, who wrote the famous *De Hominis Dignitate*) in the humanist and Neoplatonist academy of Florence in the fifteenth century. This has been elaborated by A. Koyré in *La révolution astronomique* (Paris, 1961). In 1960 Robert McNulty discovered an eyewitness account of Giordano Bruno's lectures on Copernicus at Oxford in 1583 which shows that Bruno drew heavily on Ficino's *De vita coelitus comparanda;* this is discussed by Frances A. Yates in *Giordano Bruno and the Hermetic Tradition* (London, 1964). The general subject has also been attractively discussed recently by Arthur Koestler in *The Sleepwalkers* (London, 1959), and earlier in Pauli's essay on the mystic images in Kepler's science in *Naturerklärung und Psyche* by C. G. Jung and W. Pauli (Zurich, 1952).

6. The derivation of Blake's drawing from the Renaissance studies, by Leonardo and others, of the Vitruvian proportions and mathematical harmonies of the human figure is also discussed by Sir Kenneth Clark in *The Nude* (London, 1956). It was first remarked by Sir Anthony Blunt in the *Journal of the Warburg Institute* in 1938.

7. The music of the spheres was itself a mathematical conception, which had been invented by Pythagoras in the sixth century B.C. Pythagoras taught that the distances between the heavenly bodies match the lengths of the strings that sound the different musical notes. It was deduced that the spheres that carry the heavenly bodies make music as they turn.

8. In one of the places in which Coleridge put forward this definition, the essays *On the Principles of Genial Criticism* (which Coleridge thought 'the best things he had ever written'), he traced it back to Pythagoras: 'The safest definition, then, of Beauty, as well as the oldest, is that of Pythagoras: THE REDUCTION OF MANY TO ONE.'

9. Pope was near the end of his career, and his friends Gay and Arbuthnot

were already dead, when he published these lines in 1737. (They expand a thought from Horace, and his surviving friend Swift was particularly moved by them.) Twenty-five years earlier, as a young man in *The Rape of the Lock*, Pope had pictured the mill as a happy symbol in the ritual of the coffee-table.

> For lo! the Board with Cups and Spoons is crown'd,
> The Berries crackle, and the Mill turns round.

As the eighteenth century moved on, the image of the mill became more menacing in the minds of poets, until Blake in 1804 wrote of 'dark Satanic Mills.' In part the change kept step with the progress of the Industrial Revolution, which Blake, for example, felt very sensitively. But in the main what the romantic poets feared was the new vision of nature as a machine, which Newton's great reputation had imposed. Blake meant by the Satanic Mills not a factory but the imperturbable cosmic mechanism which was now imagined to drive the planets round their orbits. Blake used the words abstract, Newtonian and Satanic with the same meaning, to describe a machinery that seemed to him opposed to organic life. (So John Constable said of a painting which he despised, 'Such things are marvellous and so is watchmaking.') Goethe, who did original work in biology, also disliked Newton's view of science; like other poets of the time, he felt that it turned the world into a clockwork. Yet at the same time religious apologists like William Paley in his *Evidences of Christianity* were using the same analogy to prove that the world, like a clock, must have an intelligent designer. Thus the symbol of the clockwork, and (as T. S. Ashton has pointed out) a new sense of time in general, were critical in the thought of those who lived through the Industrial Revolution.

10. A literal translation is:

> If ever I say to the present moment,
> "Please stay! You are so beautiful!"
> Then you may cast me in fetters,
> Then I will gladly perish!
> Then let the death bell toll,
> Then you are released from your service.
> Let the clock stop, let the hand fall,
> Let time be at an end for me!

The greatest satire of the First World War, Karl Kraus's *Die Letzten Tage der Menschheit*, contains a moving echo of these lines, which bears on what I have written in the preceding note. In one poem Kraus describes the machine-made murders of modern war as observed by a man *Mit der Uhr in der Hand*—that is, watch in hand. I quote two verses.

> Dort ist ein Mörser. Ihm entrinnt der arme Mann,
> der ihn erfand. Er schützt sich in dem Graben.

Weil Zwerge Riesen überwältigt haben,
seht her, die Uhr die Zeit zum Stehen bringen kann!

Wie viel war's an der Zeit, als jenes jetzt geschah?
Schlecht sieht das Aug, das giftige Gase beizen.
Doch hört das Ohr, die Uhr schlug eben dreizehn.
Unsichtig Wetter kommt, der Untergang ist nah.

A literal translation is:

There is a mortar. From it escapes the wretched man
who invented it. He takes refuge in the trench.
Because dwarfs have overpowered giants,
behold, the clock can bring time to a stop!

What time was it when this was happening now?
The eye sees poorly that is etched by poison gases.
But the ear hears, the clock just struck thirteen.
Misty weather is coming, destruction is near.

The same image of the ticking clockwork haunted me when I visited refugee
camps after another war, in 1947; and I wrote,

The voice of God that spoke and struck
Was the cuckoo in the clock.
The exiles in the garden heard
The engine tremble in the bird,
Sobbing throat and iron bill:
Time on his springy wheel stood still.

Time began and time runs down.
The voices in the garden drown.
No God from his machine unhands
The exile with a mouth of sand.
The clockwork cuckoo on the hill,
Abrupt and wheeling, stoops to kill.

11. This verse comes from the poem 'Vacillation,' and I have quoted it as
Yeats first printed it, for example in *The Winding Stair and other poems*
In his *Collected Poems* soon after, Yeats left out the word *or* in the second
line. No doubt the change improves the meter; but since I am here concerned with the contrast between the two images in Yeats's mind, I have
given his original text.

two

The Habit of Truth

IN 'The Creative Mind' I set out to show that there exists a single creative activity, which is displayed alike in the arts and in the sciences. It is wrong to think of science as a mechanical record of facts, and it is wrong to think of the arts as remote and private fancies. What makes each human, what makes them universal, is the stamp of the creative mind.

I found the act of creation to lie in the discovery of a hidden likeness. The scientist or the artist takes two facts or experiences which are separate; he finds in them a likeness which had not been seen before; and he creates a unity by showing the likeness.

The act of creation is therefore original; but it does not stop with its originator. The work of art or of science is universal because each of us re-creates it. We are moved by the poem, we follow the theorem because in them we discover again and seize the likeness which their creator first seized. The act of appreciation re-enacts the act of creation, and we are (each of us) actors, we are interpreters of it.

My examples were drawn from physics and from poetry, because these happen to be the works of man which I know best. But what is great in these is common to all great works. In the museum at Cracow there is a painting by Leonardo da Vinci called *The Lady with a Stoat* : it shows a girl holding a stoat in her arms. The girl was probably a mistress of Lodovico Sforza, the usurper of Milan, at whose court Leonardo lived from about 1482 to 1499, amid the violence and intrigue which all his life drew him and repelled him together. The stoat was an emblem of Lodovico Sforza, and is probably also a pun on the girl's name. And in a sense the

whole picture is a pun, if I may borrow for the word the tragic
intensity which Coleridge found in the puns of Shakespeare.
Leonardo has matched the stoat in the girl. In the skull under
the long brow, in the lucid eyes, in the stately, brutal, beautiful and stupid head of the girl, he has re-discovered the animal nature; and done so without malice, almost as a matter
of fact. The very carriage of the girl and the stoat, the gesture
of the hand and the claw, explore the character with the ana-
tomy. As we look, the emblematic likeness springs as freshly
in our minds as it did in Leonardo's when he looked at the
girl and asked her to turn her head. *The Lady with a Stoat*
is as much a research into man and animal, and a creation of
unity, as is Darwin's *Origin of Species*.[1]

2

So much may be granted; and yet, where is it to stop? The
creative act is alike in art and in science; but it cannot be
identical in the two; there must be a difference as well as
a likeness. For example, the artist in his creation surely has
open to him a dimension of freedom which is closed to the
scientist. I have insisted that the scientist does not merely
record the facts; but he must conform to the facts. The sanction of truth is an exact boundary which encloses him, in a
way in which it does not constrain the poet or the painter.
Shakespeare can make Romeo say things about the look of
Juliet which, although they are revealing, are certainly not
true in fact.

O she doth teach the Torches to burne bright.

But soft, what light through yonder window breaks?
It is the East, and *Juliet* is the Sunne.

And Shakespeare himself is aware that these statements differ
from those made by exact observers. For he exploits the

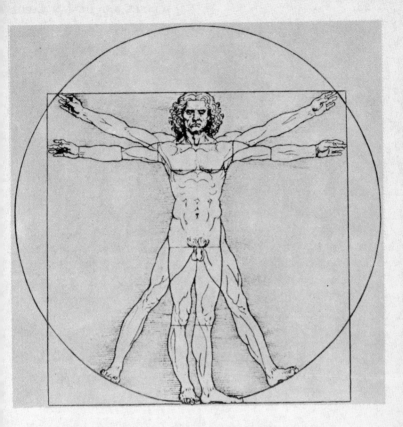

The Proportions of the Human Figure (after Vitruvius) *by Leonardo da Vinci*

difference deliberately for a new poetic effect in the sonnet
which begins, tartly,

> My Mistres eyes are nothing like the Sunne.

This takes its point and pungency from being unpoetic.
Shakespeare designedly in this sonnet plays the finicking
scientist straightfaced—

> Currall is farre more red, then her lips red,
> If snow be white, why then her brests are dun

—in order to say at last, overwhelmingly, that even in plain
fact his love is incomparable. No doubt Shakespeare would
have been willing to argue in other places that the poetic
image can be called true: the parable of the Prodigal Son is
true in some sense, and so is the pursuit of Orestes by the
furies, and the imagery of *Romeo and Juliet* itself. But the
sonnet proves that Shakespeare did not think this meaning
of truth to be the same as that which he met in Holinshed's
Chronicles and William Gilbert's *De Magnete,* and which
now dogs the writer of a thesis on electronic networks.

We cannot shirk the historic question, What is truth? On
the contrary: the civilization we take pride in took a new
strength on the day the question was asked. It took its greatest strength later from Renaissance men like Leonardo, in
whom truth to fact became a passion. The sanction of experienced fact as a face of truth is a profound subject, and the
mainspring which has moved our civilization since the
Renaissance.

3

Those who have gone out to climb in the Himalayas have
brought back, besides the dubious tracks of Abominable
Snowmen, a more revealing model of truth. It is contained

in the story which they tell of their first sight of some inaccessible and rarely seen mountain. The western climbers, at home with compass and map projection, can match this view of the mountain with another view that they have seen years ago. But to the native climbers with them, each face is a separate picture and puzzle. The natives may know another face of the mountain, and this face too, better than the strangers; and yet they have no way of fitting the two faces together. Eric Shipton describes this division in the account of his reconnaissance for the new route to Everest, on which the later ascent in 1953 was based. Here is Shipton moving up to a view of Everest from the south, which is new to him, but which his leading Sherpa, Angtarkay, had known in childhood :

> On the morning of the 27th we turned into the Lobujya Khola, the valley which contains the Khombu Glacier (which flows from the south and south-west side of Everest). As we climbed into the valley we saw at its head the line of the main watershed. I recognized immediately the peaks and saddles so familiar to us from the Rongbuk (the north) side: Pumori, Lingtren, the Lho La, the North Peak and the west shoulder of Everest. It is curious that Angtarkay, who knew these features as well as I did from the other side and had spent many years of his boyhood grazing yaks in this valley, had never recognized them as the same; nor did he do so now until I pointed them out to him.

The leading Sherpa knows the features of Everest from the north as well as Shipton does. And unlike Shipton, he also knows them from the south, for he spent years in this valley. Yet he has never put the two together in his head. It is the inquisitive stranger who points out the mountains which flank Everest. The Sherpa then recognizes the shape of a peak here and of another there. The parts begin to fit together; the puzzled man's mind begins to build a map; and suddenly the pieces are snug, the map will turn around, and the two faces of the mountain are both Everest. Other expeditions in other places have told of the delight of the native climbers at such a recognition.

All acts of recognition are of this kind. The girl met on the beach, the man known long ago, puzzle us for a moment and then fall into place; the new face fits on to and enlarges the old. We are used to make these connections in time; and like the climbers on Everest we make them also in space. If we did not, our minds would contain only a clutter of isolated experiences. By making such connections we find in our experiences the maps of things.

There is no other evidence for the existence of things. We see the left profile of a man and we see the right profile; we never see them together. What are our grounds for thinking that they belong to one man? What are the grounds for thinking that there is such a thing as the one man at all? By the canons of classical logic there are no grounds: no one can deduce the man. We infer him from his profiles as we infer that the evening star and the morning star are both the planet Venus; because it makes two experiences cohere, and experience proves it to be consistent. Profiles and full face, back and front, the parts build a round whole, not only by sight but by the exploration of the touch and the ear, and the stethoscope and the X-ray tube and all our elaborations of inference. Watch a child's eyes and fingers together discover that the outside and the inside of a cup hang together. Watch a man who was born blind, and who can now see, re-building the touched world by sight; and never again think that the existence of a thing leaps of itself into the mind, immediate and whole. We know the thing only by mapping and joining our experiences of its aspects.[2]

4

The discovery of things is made in three steps. At the first step there are only the separate data of the senses: we see the head of the penny, we see the tail. It would be mere pomp to use words as profound as true and false at this simple step. What we see is either so or is not so. Where no other judgment can be made, no more subtle words are in place.

At the second step we put the head and the tail together. We see that it makes sense to treat them as one thing. And the thing is the coherence of its parts in our experience.

The human mind does not stop there. The animal can go as far as this: an ape will learn to recognize a cup whenever and however he sees it, and will know what to do with it. But all that has been learned about apes underlines that they find it hard to think about the cup when it is not in sight, and to imagine its use then. The human mind has a way of keeping the cup or the penny in mind. This is the third step: to have a symbol or a name for the whole penny. For us the thing has a name, and in a sense is the name: the name or symbol remains present, and the mind works with it, when the thing is absent. By contrast, one of the difficulties which the Sherpas have in seeing Everest is that the mountain goes by different names in different valleys.[3]

The words true and false have their place at the latter steps, when the data of the senses have been put together to make a thing which is held in the mind. Only then is it meaningful to ask whether what we think about the thing is true. That is, we can now deduce how the thing should behave, and see whether it does so. If this is really one mountain, we say, then the bearing of that landmark should be due east; and we check it. If this is a penny, then it should be sensible to the touch. This is how Macbeth tests the thing he is thinking about and seems to see.

> Is this a Dagger, which I see before me,
> The Handle toward my Hand? Come, let me clutch thee—

Macbeth is using the empirical method: the thing is to be tested by its behavior.

> Come, let me clutch thee:
> I haue thee not, and yet I see thee still.
> Art thou not fatall Vision, sensible
> To feeling, as to sight? or art thou but
> A Dagger of the Minde, a false Creation?

'A Dagger of the Minde, a false Creation;' both the word false and the word Creation are exact. What the human mind makes of the sense data, and thinks about, is always a created thing. The construction is true or false by the test of its behavior. We have constructed the thing from the data; we now deduce how the thing should behave; and if it does not, then our construction was false. What was false was not the sense data but our intrepretation of them : we constructed a hallucination.

5

I have described so far how we think about things. The view I have put forward also looks beyond things, to the laws and concepts which make up science. This is the real reach of this view: that the three steps by which man constructs and names a mountain are also the steps by which he makes a theory.

Recall the example of the work of Kepler and of Newton; the steps are there to be re-traced. The first step is the collection of data: here, astronomical observations. Next comes the creative step which Kepler took, which finds an order in the data by exploring likenesses. Here the order, the unity, is the three laws by which Kepler described the orbit, not of this planet or of that, but simply of a planet.

Kepler's laws, however, put forward no central concept; and the third step is to create this concept. Newton took this step when, at the center of astronomy, he put a single activity of the universe: the concept of gravitation.

There is of course no such thing as gravitation, sensible to the touch. It is neither seen nor heard; and if it seems to be felt, this now appears to be a quirk of space and time.[4] Yet the concept of gravitation was real and true. It was constructed from the data by the same steps which fuse two views of Everest into one mountain, or many conversations into one man. And the concept is tested as we test the man, by its behavior: it must be in character. Newton was doing this in his garden in 1666 when he computed the force which holds the moon in her orbit; like Macbeth, he was testing the creation of the mind.

The creation was a concept—a connected set of concepts. There was the concept of a universal gravitation, reaching beyond the tree tops and the air to the ends of space. There was the concept of other universal forces in space, which try to pull the moon away as a whirling stone pulls away from its string. And there was the concept which put an end to the four elements of Aristotle: the concept of mass, alike in the apple and the earth and the moon, in all earthly and all heavenly bodies.[5]

All these are real creations: they find a unity in what seemed unlike. They are symbols; they do not exist without the creation. Solid as it seems, there is no such thing as mass; as Newton ruefully found, it cannot be defined. We experience mass only as the behavior of bodies, and it is a single concept only because they behave consistently.

Indeed, the concept of mass is a peculiarly apt example. For in the physics of Newton there are two concepts of mass, which are distinct. One is the inertial mass of a body—that which must be overcome when it is thrown. The other is the gravitational mass of the same body—that which must be overcome when it is lifted from the ground. Newton knew, of course, that the inertial mass is equal to the gravitational mass; but why are they equal, why should they be the same single mass? This is a question which Einstein asked; and in order to answer it he built in 1915 the whole theory of

General Relativity. Only in that theory were the two faces of mass made one, and made the unity which is the single concept, mass.

This sequence is characteristic of science. It begins with a set of appearances. It organizes these into laws. And at the center of the laws it finds a knot, a point at which several laws cross: a symbol which gives unity to the laws themselves. Mass, time, magnetic moment, the unconscious: we have grown up with these symbolic concepts, so that we are startled to be told that man had once to create them for himself. He had indeed, and he has; for mass is not an intuition in the muscle, and time is not bought ready-made at the watchmaker's.

And we test the concept, as we test the thing, by its implications. That is, when the concept has been built up from some experiences, we reason what behavior in other experiences should logically flow from it. If we find this behavior, we go on holding the concept as it is. If we do not find the behavior which the concept logically implies, then we must go back and correct it. In this way logic and experiment are locked together in the scientific method, in a constant to and fro in which each follows the other.

6

This view of the scientific method is not shared by all those who have thought about it. There are two schools of philosophy which are suspicious of conceptual thinking, and want to replace it wholly by the manipulation of facts. One is that offshoot of the English empiricist tradition which goes through Bertrand Russell to Wittgenstein and the logical positivists. This school holds that a rigorous description of all nature can be pieced together, like a gigantic tinker-toy, out of small units of fact, each of which can be separately verified to be so. The other is the school founded by Ernst

Mach in Austria, and led more recently by Percy Bridgman in America, which holds that science is strictly an account of operations and their results. This behaviorist school would like to discard all models of nature, and confine itself always to saying that if we do *this*, we get a larger measurement than if we do *that*.

These accounts of science seem to me mistaken, on two counts. First, they fly in the face of the historical evidence. Since Greek times and before, lucid thinkers and indeed all men have used such words as space and mass and light. They have not asked either Russell's or Bridgman's leave, yet what they have done with the words belongs to the glories of science as well as philosophy; and it is late in the day to forbid them this language.

And second, both schools fly in the face of the contemporary evidence. We have good grounds to believe, from studies of animals and men, that thinking as we understand it is made possible only by the use of names or symbols. Other animals than man have languages, in Bridgman's sense; for example, bees signal to one another where to go in order to find nectar.[6] Bernard de Mandeville, who wrote *The Fable of the Bees* in an eighteenth-century parable, would have thought this the height of rational behavior. But no active scientist sees it so, because he knows that science is not something which insects or machines can do. What makes it different is a creative process, the exploration of likenesses; and this has sadly tiptoed out of the mechanical worlds of the positivists and the operationalists, and left them empty.

The world which the human mind knows and explores does not survive if it is empied of thought. And thought does not survive without symbolic concepts. The symbol and the metaphor are as necessary to science as to poetry. We are as helpless today to define mass, fundamentally, as Newton was. But we do not therefore think, and neither did he, that the equations which contain mass as an unknown are mere rules of thumb. If we had been content with that view, we should never have learned to turn mass into energy. In forming a concept of mass, in speaking the word, we begin a process of experiment and correction which is the creative search for truth.[7]

In the village in which I live there is a pleasant doctor who is a little deaf. He is not shy about it and he wears a hearing aid. My young daughter has known him and his aid since she was a baby. When at the age of two she first met another man who was wearing a hearing aid, she simply said, 'That man is a doctor.' Of course she was mistaken. Yet if both men had worn not hearing aids but stethoscopes, we should have been delighted by her generalization. Even then she would have had little idea of what a doctor does, and less of what he is. But she would have been then, and to me she was even while she was mistaken, on the path to human knowledge which goes by way of the making and correcting of concepts.

7

I should be unjust if I did not grant that the positivist and operational schools of philosophy have had reason to be wary of the appeal to concepts. Russell and Bridgman shied away from the concept because it has a bad record, which still befuddles its use. Historically, concepts have commonly been set up as absolute and inborn notions, like the space and time which Kant believed to be ready-made in the mind. The view that our concepts are built up from experience, and have constantly to be tested and corrected in experience, is not classical. The classical view has been that concepts are not accessible to empirical tests. How many people understand, even today, that the concepts of science are neither absolute nor everlasting? And beyond the field of science, in society, in personality, above all in ethics, how many people will allow the sanction of experienced fact? The common view remains the classical view, that the concepts of value—justice and honour, dignity and tolerance—have an inwardness which is inaccessible to experience.

The roots of this error go down into the closed logic of the Middle Ages. The characteristic and distinguished example is the method of St. Thomas Aquinas. The physics which was current for three centuries before the Scientific Revolution derived from Aristotle by way of Arab scholars, and had been formed into a system by Aquinas. But it did not share the test of truth of modern physics. Between the years 1256 and 1259 Aquinas held about 250 discussion classes, all on the subject of truth. Each class lasted two days. The questions discussed belong to a world of discourse which simply has no common frontier with ours. They are such questions as 'Is God's knowledge the cause of things?' 'Is the Book of Life the same as predestination?' 'Do angels know the future?'

I do not dismiss these as merely fanciful questions, any more than I regard *Tamburlaine* and *The Marriage of Heaven and Hell* as fantasies. Yet it is plain that they have no bearing on matters of truth and falsehood as we understand them, inductively. These debates are scholastic exercises in absolute logic. They begin from concepts which are held to be fixed absolutely; they then proceed by deduction; and what is found in this way is subject to no further test. The deductions are true because the first concepts were true: that is the scholastic system. It is also the logic of Aristotle. Unhappily, it makes poor physics, for there the gap between the intuitive and the corrected concept is gaping.

8

Modern science also began by hankering after purely deductive systems. Its first teacher, of course, was Euclid. One of its historic moments was the conversion of Thomas Hobbes, some time between 1629 and 1631:

He was 40 yeares old before he looked on geometry; which happened accidentally. Being in a gentleman's library, Euclid's Elements lay open, and 'twas the 47 El. libri I. He read the proposition. 'By G—,' sayd he, (He would now and then sweare, by way of emphasis) 'By G—,' sayd he, 'this is impossible!' So he reads the demonstration of it, which referred him back to such a proposition; which proposition he read. That referred him back to another, which he also read. *Et sic deinceps,* that at last he was demonstratively convinced of the trueth. This made him in love with geometry.

This account was written by John Aubrey, who was Hobbes's friend. Aubrey of course assumes that everyone knows which is the 47th proposition in the first book of Euclid; if we do not, we miss the explosive charge in the story. For the 47th proposition is the theorem of Pythagoras about the squares on the sides of a right-angled triangle—the most famous theorem of antiquity, for which Pythagoras is said to have sacrificed a hundred oxen to the Muses in thanks. Hobbes, in an age which knew the theorems by their numbers, at forty years old did not know the content of this; and when he learned it, it changed his life.

From then on Hobbes became a pioneer of the deductive method in science. In his time, his innovation was necessary; but soon the movement of science left it behind. For when Hobbes took over the deductive method, he took also Euclid's notion that we know intuitively what points are, what an angle is, what we mean by parallel. The concepts and the axioms were supposed to be simply self-evident, in geometry or in the physical world.

Science has not stopped there since Hobbes, but such subjects as ethics have. In Hobbes's lifetime Spinoza presented his *Ethics 'ordine geometrico demonstrate,'* proved in geometrical order. The book begins in the Euclidean manner with eight definitions and seven axioms. This is a modest apparatus with which to attack the universe, for even Euclid's geometry of the plane needs more than twenty axioms. But Spinoza tackles it bravely and indeed profoundly, and it is not his fault that after a time we come to feel that we are standing still. The geometrical system of ethics has exhausted its discoveries. It no longer says anything new; and worse, it can learn nothing new.

Here is the heart of the difference between the two ways in which we order our lives. Both ways hinge on central concepts. In both we reason from the central concepts to the consequences which flow from them. But here the two ways divide. In the fields of ethics, of conduct and of values, we think as Aquinas and Spinoza thought: that our concepts must remain unchangeable because they are either inspired or self-evident. In the field of science, four hundred years of adventure have taught us that the rational method is more subtle than this, and that concepts are its most subtle creations. A hundred and fifty years ago Gauss and others proved that the axioms of Euclid are neither self-evident nor necessarily exact in our world. Much of physics since then, for instance in Relativity, has been the re-making of a more delicate and a more exciting concept of space. The need to do so has sprung from the facts; and yet, how the new concepts have outraged our self-evident notions of how a well-mannered space *ought* to behave! Quantum mechanics has been a constant scandal because it has said that the world of the small scale does not behave entirely like a copy of the man-sized world. Jonathan Swift in *Gulliver's Travels* had remarked something like this back in 1726, and it ought no longer to shock us; but of course Swift was a scandal too in his day.

9

Is it true that the concepts of science and those of ethics and values belong to different worlds? Is the world of *what is* subject to test, and is the world of *what ought to be* subject to no test? I do not believe so. Such concepts as justice, humanity and the full life have not remained fixed in the last four hundred years, whatever churchmen and philosophers may pretend. In their modern sense they did not exist when Aquinas wrote; they do not exist now in civilizations which disregard the physical fact. And here I do not mean only the scientific fact. The tradition of the Renaissance is of a piece, in art and in science, in believing that the physical world is a source of knowledge. The poet as much as the biologist now believes that life speaks to him through the senses. But this was not always so: Paolo Veronese was reproved by the Inquisition in 1573 for putting the everyday world into a sacred painting.[8] And it is not so everywhere now: the ancient civilizations of the East still reject the senses as a source of knowledge, and this is as patent in their formal poetry and their passionless painting as in their science.

By contrast, the sanction of experienced fact has changed and shaped all the concepts of men who have felt the Scientific Revolution. A civilization cannot hold its activities apart, or put on science like a suit of clothes—a workday suit which is not good enough for Sundays. The intellectual study of perspective in the Renaissance chimes with the rise of sensuous painting. And the distaste of painters for naturalism for fifty years now is surely related to the new structure which scientists have struggled to find in nature in the same time. A civilization is bound up with one way of experiencing life. And ours can no more keep its concepts than its wars apart in pigeonholes.

All this is plain once it is seen that science also is a system of concepts: the place of experience is to test and correct the concept. The test is, Will the concept work? Does it give an unforced unity to the experience of men? Does the concept make life orderly, not by edict but in fact?

Men have insisted on carrying this test into the systems of society and of conduct. What else cost Charles I his head in 1649? And what brought Charles II back in 1660, yet at last exiled his family for Dutchmen and Germans? Not the high

talk about the divine right of kings, and not the Bill of
Rights, but their test in experience. England would have
been willing to live by either concept, as it has been willing
to live by Newton or by Einstein: it chose the one which
made society work of itself, without constraint.

Since then society has evolved a sequence of central concepts each of which was at one time thought to make it work
of itself, and each of which has had to be corrected to the
next. There was the early eighteenth-century concept of self-
interest, in Mandeville and others; then came enlightened
self-interest; then the greatest happiness of the greatest number; utility; the labor theory of value; and thence its
expression either in the welfare state or in the classless society.
Men have never treated any one of these concepts as the last,
and they do not mean to do so now. What has driven them,
what drives them, is the refusal to acknowledge the concept
as either an edict or self-evident. Does this really work, they
ask, without force, without corruption, and without another
arbitrary superstructure of laws which do not derive from the
central concept? Do its consequences fit our experience; do
men in such a society live so or not so? This is the simple
but profound test of fact by which we have come to judge
the large words of the makers of states and systems. We see it
cogently in the Declaration of Independence, which begins in
the round Euclidean manner 'We hold these truths to be
self-evident,' [9] but which takes the justification for its action
at last from 'a long train of abuses and usurpations': the colonial system had failed to make a workable society.

10

One example among others points the modern lesson. When
Warren Hastings was impeached in 1786 for his high-handed
rule in India, he had two grievances on his side. One was that
the violence and corruption of which he was accused were in

any case commonplace throughout Indian society then. The other was that some of his accusers (and chiefly Edmund Burke) were not free from a corrupt interest in Indian affairs themselves. Hastings was acquitted, but not on these grounds, for they missed the difference between India then and Europe. Man as man had a different value in the two continents. The Renaissance had made the difference; and England with her dissenters had been evolving the new value for two hundred years, always by the downright test of making a stable and self-correcting society. The conduct of Warren Hastings was to be judged by the same aim, and by no other; the standards of lesser societies ruled by conquerors, the motives of lesser men, had no bearing on it.[10]

The cultures of the East still differ from ours as they did then. They still belittle man as individual man. Under this runs an indifference to the world of the senses, of which the indifference to experienced fact is one face. Anyone who has worked in the East knows how hard it is there to get an answer to a question of fact. When I had to study the casualties from the atomic bombs in Japan at the end of the war, I was dogged and perplexed by this difficulty. The man I asked, whatever man one asks, does not really understand what one wants to know: or rather, he does not understand that one wants to know. He wants to do what is fitting, he is not unwilling to be candid, but at bottom he does not know the facts because they are not his language. These cultures of the East have remained fixed because they lack the language and the very habit of fact.

To us, the habit of simple truth to experience has been the mover of civilization. The last war showed starkly what happens to our societies and to us as men when this habit is broken. The German occupation of France forced on the people of France a split in the conduct of each man: a code of truth to his fellows, and a code of deceit to the conquerors. This was a heroic division, more difficult to sustain than we can know, and for which the world has still to thank Frenchmen. Yet those who lived in that division will never wholly recover from it, and the habits of distrust and withdrawal which it created will long hamstring the free life of France and of Europe.[11]

This is the grave indictment of every state in which men are cautious of speaking out to any man they meet. The decay of the habit of truth is damaging to those who must fear to speak. But how much more destructive, how degrading it is to the loud-mouthed conquerors! The people whom their conquests really sapped were the Nazis themselves. Picture the state of German thought when Werner Heisenberg was criticized by the S.S., and had to ask Himmler to support his scientific standing. Heisenberg had won the Nobel prize at the age of thirty; his principle of uncertainty is one of the two or three deep concepts which science has found in this century; and he was trying to warn Germans that they must not dismiss such discoveries as Relativity because they disliked the author. Yet Himmler, who had been a schoolmaster, took months of petty inquiry (someone in his family knew Heisenberg) before he authorized of all people Heydrich to protect Heisenberg. His letter to Heydrich is a paper monument to what happens to the creative mind in a society without truth. For Himmler writes that he has heard that Heisenberg is good enough to be earmarked later for his own Academy for *Welteislehre*. This was an Academy which Himmler proposed to devote to the conviction which he either shared with or imposed on his scientific yes-men, that the stars are made of ice.

This nonsense of course is like the nonsense that Germans were taught to credit about the human races. The state of mind, the state of society is of a piece. When we discard the test of fact in what a star is, we discard it in what a man is. A society holds together by the respect which man gives to man; it fails in fact, it falls apart into groups of fear and

THE HABIT OF TRUTH

power, when its concept of man is false. We find the drive which makes a society stable at last in the search for what makes us men. This is a search which never ends: to end it is to freeze the concept of man in a caricature beyond correction, as the societies of caste and master-race have done. In the knowledge of man as in that of nature, the habit of truth to experienced fact will not let our concepts alone. This is what destroyed the empires of Himmler and of Warren Hastings. When Hastings stood his trial, William Wilber- force was rousing England to put an end to the trade in slaves. He had at bottom only one ground: that dark men are men. A century and more of scientific habit by then had made his fellows find that true, and find Hastings not so much a tyrant as a cheat.

11

There have always been two ways of looking for truth. One is to find concepts which are beyond challenge, because they are held by faith or by authority or the conviction that they are self-evident. This is the mystic submission to truth which the East has chosen, and which dominated the axiomatic thought of the scholars of the Middle Ages. So St. Thomas Aquinas holds that faith is a higher guide to truth than knowledge is: the master of medieval science puts science firmly into second place.

But long before Aquinas wrote, Peter Abelard had already challenged the whole notion that there are concepts which can only be felt by faith or authority. All truth, even the highest, is accessible to test, said Abelard: 'By doubting we are led to inquire, and by inquiry we perceive the truth.' The words might have been written five hundred years later by Descartes, and could have been a recipe for the Scientific Revolution. We have the same dissent from authority in the Reformation; for in effect what Luther said in 1517 was that

we may appeal to a demonstrable work of God, the Bible, to
override any established authority. The Scientific Revolution begins when Copernicus implied the bolder proposition
that there is another work of God to which we may appeal
even beyond this: the great work of nature. No absolute statement is allowed to be out of reach of the test, that its conse-
quence must conform to the facts of nature.[12]

The habit of testing and correcting the concept by its consequences in experience has been the spring within the movement of our civilization ever since. In science and in art and
in self-knowledge we explore and move constantly by turning to the world of sense to ask, Is this so? This is the habit
of truth, always minute yet always urgent, which for four
hundred years has entered every action of ours; and has made
our society and the value it sets on man, as surely as it has
made the linotype machine and the scout knife, and *King
Lear* and the *Origin of Species* and Leonardo's *Lady with a
Stoat*.

NOTES

1. The example of Leonardo is particularly relevant to the topic of this
essay, because he was a pioneer in disregarding the large theories which
dominated science in the Middle Ages, and in turning instead to the exact
test of fact. For example, Leonardo was one of the first anatomists to
dissect human corpses, and to draw what he saw and not what Galen had
laid down he would see. He was one of several Renaissance painters who
were inspired by the detail in nature, and he was unique in taking this
discovery from the artist's studio into the scientist's study. By making
scientists aware that the detail of nature is significant, and is the test of a
theory, Leonardo helped to establish the scientific method with which this
essay deals. I have discussed his fundamental achievement and the Renaissance outlook in an essay, 'The Creative Process,' in the *Scientific American*
of September 1958, and again in the same journal in June 1963; and in the
opening chapters of *The Western Intellectual Tradition* (Harper Torchbooks,
1962).

2. The argument here is a scientific formulation and extension of the position which philosophers call phenomenalism. Usually philosophers discuss only a primitive form of phenomenalism. There is now good evidence for the argument I have put forward, both from studies in the psychology of perception and, oddly, from work on machines which 'learn.' For example, Dr. A. M. Utley has built a machine at the National Physical Laboratory which learns to recognize and to associate differently shaped shadows which the same object casts in different positions—just as chicks learn to recognize the shadow of a hawk in any position.

3. It has long been familiar that language and other symbolisms play a central part in thought. Recently the importance of the metaphorical element in any symbolism has been stressed by philosophers—for example, by Susanne K. Langer in *Philosophy in a New Key* (Harvard, 1942). In science the concepts have to fit into the larger symbolism of some model of nature, which (as I remarked of Kepler and of Rutherford) has its own metaphorical associations.

. General Relativity treats all accelerated motions alike, and disregards the different forces which were postulated by Newton to produce them. On this view, then, a massive body is not the source of a gravitational force of attraction, but is simply the center of a geometrical configuration which guides other bodies towards it—a sort of hollow in space-time.

. The evolution of these concepts step by step from Aristotle to Robert Hooke and Newton has been traced in the classical studies of Pierre Duhem.

. Bees and other animals use language to inform and instruct, that is, as a means of public communication. Man alone has developed the private use of language to manipulate ideas inside his head. In this as in other gifts, man is a double creature: he is the social solitary, who needs to be sustained by his fellows yet to think alone. I have described this duality in detail in the first of two Frank Gerstein lectures which I gave at York University, *The Imaginative Mind in Art and in Science* (Toronto, 1964).

. The positivist and the operational schools share the view that scientific concepts are purely logical constructions—a view which Bertrand Russell stated thus in 1918: 'The supreme maxim in scientific philosophizing is this: *Wherever possible, logical constructions are to be substituted for inferred entities.*' On this view the word electron, for example, can be explicitly defined in terms of known observations; and every sentence which contains the word electron can in principle be translated without loss of meaning into a sentence which contains only observations. As long ago as 1929, shortly before his death at the age of twenty-six, F. P. Ramsey showed that this view is untenable. He did this by constructing a logical example which proves, in effect, that a system which contains only logical constructions cannot discover or accommodate new relations. The matter has since been admirably discussed, and simpler examples have been constructed, by R. B. Braithwaite in *Scientific Explanation* (Harper Torchbooks, 1960). It is on the basis of these examples that I say that a system which defines mass by logical construction leaves no scope for the discovery that mass is equivalent to energy.

8. Sir Kenneth Clark in *The Nude* points to the influence which the Council of Trent had in discouraging the entry of sensuous and human details into sacred pictures. The Council met between 1545 and 1563 to settle the doctrines of the Roman Catholic Church in opposition to the Reformation.

9. Indeed, in Thomas Jefferson's original draft the Declaration of Independence began with an even larger claim 'We hold these truths to be sacred and undeniable.' The substitution of the word self-evident, with its scientific overtones, is thought to have been made by Benjamin Franklin. But of course the word self-evident begs as many questions as do the words sacred and undeniable.

10. The crux of principle was put clearly by the Whig leader, Charles James Fox. As he said, there was only one issue in the impeachment: whether India should be governed 'by those laws which are to be found in Europe, Africa, and Asia—that are found amongst all mankind, those principles of equity and humanity, implanted in our hearts, which have their existence in the feelings of mankind that are capable of judging.'

11. The Catholic writer François Mauriac has returned to this theme recently, in a series of attacks on the violence in the French army and on the gangster morality of what he calls the high-class underworld of the rich. He recalls a prophetic phrase by the poet Rimbaud, '*Voici venu le temps des assassins,*' and comments: 'This epoch of assassins did not come suddenly. It formed slowly during the long struggles that since 1914 have bloodied Europe. Foreign wars helped to prepare it, but less than the civil wars and the Resistance movement, in which the duty of each partisan was to make a virtue of crime.'

12. This was said explicitly by Galileo when he had to defend the system of Copernicus in 1615, in the *Letter to the Grand Duchess of Tuscany, Concerning the Use of Biblical Quotations in Matters of Science*. For example Galileo writes: 'Nature is inexorable and immutable; she never transgresses the laws imposed upon her, or cares a whit whether her abstruse reasons and methods of operation are understandable to men. For that reason it appears that nothing physical which sense-experience sets before our eyes, or which necessary demonstrations prove to us, ought to be called in question (much less condemned) upon the testimony of biblical passages which may have some different meaning beneath their words.' In arguing in this way Galileo leans heavily on the support of St. Augustine.

three

The Sense of
Human Dignity

THE subject of this book is the evolution of contemporary values. My theme is that the values which we accept today as permanent and often as self-evident have grown out of the Renaissance and the Scientific Revolution. The arts and the sciences have changed the values of the Middle Ages; and this change has been an enrichment, moving towards what makes us more deeply human.

This theme plainly outrages a widely held view of what science does. If, as many think, science only compiles an endless dictionary of facts, then it must be as neutral (and neuter) as a machine is; it cannot bear on human values. But of course science is not a giant dictionary, any more than literature is; both are served by, they do not serve, the makers of their dictionaries. 'The Creative Mind' had this strenuous task, to show that science is a creative activity. In the act of creation, a man brings together two facets of reality and, by discovering a likeness between them, suddenly makes them one. This act is the same in Leonardo, in Keats and in Einstein. And the spectator who is moved by the finished work of art or the scientific theory re-lives the same discovery; his appreciation also is a re-creation.[1]

Yet when it has been granted that science and art both find hidden likenesses and order in what seemed unlike, there remains a doubt. Is there not this difference between them, that the likenesses which a science finds have to conform to a sanction of fact from which the arts are free? Must not science be true?

'The Habit of Truth' asked the historic question, What is truth? It set out, of course, to distinguish what is true, less

from what is simply false (which seldom puzzles us) than
from what is illusory: the hallucination of an ill-grounded or
a disordered belief. My method derived from the tradition
of pragmatism which, since William James advanced it about
1890 (and Charles Peirce before that), has been the most
original philosophical thought in America. It took for its
model of truth the reality of things. How do we come to believe that there is such a thing as Everest? For we do not see
the thing in itself; only an aspect or an effect of it reaches
us. Yet we recognize the thing as one, because this gives
order to its aspects; the thing makes a unity of the different
effects by which it enters our world.

I do not think that truth becomes more primitive if we
pursue it to simpler facts. For no fact in the world is instant,
infinitesimal and ultimate, a single mark. There are, I hold,
no atomic facts. In the language of science, every fact is a
field—a crisscross of implications, those that lead to it and
those that lead from it.

Truth in science is like Everest, an ordering of the facts.
We organize our experience in patterns which, formalized,
make the network of scientific laws. But science does not stop
at the formulation of laws; we none of us do, and none of us,
in public with his work or in private with his conscience, lives
by following a schedule of laws. We condense the laws around
concepts. So science takes its coherence, its intellectual and
imaginative strength together, from the concepts at which
its laws cross, like knots in a mesh. Gravitation, mass and
energy, evolution, enzymes, the gene and the unconscious—
these are the bold creations of science, the strong invisible
skeleton on which it articulates the movements of the world.

Science is indeed a truthful activity. And whether we look
at facts, at things or at concepts, we cannot disentangle
truth from meaning—that is, from an inner order.[2] Truth
therefore is not different in science and in the arts; the
facts of the heart, the bases of personality, are merely more

difficult to communicate. Truth to fact is the same habit in both, and has the same importance for both, because facts are the only raw material from which we can derive a change of mind. In science, the appeal to fact is the exploration of the concept in its logical consequences. In the arts, the emotional facts fix the limits of experience which can be shared in their language.

2

I have recalled the apparatus which I have previously set up in order that I may now use it to examine the values by which we live. Some people think that these values are inborn as the sense of sight is, and they treat any heresy as an affliction which the sufferer would not have contracted had his habits been cleaner. Others accept the notions of value as absolute edicts which we must indeed learn, and if possible learn to like, but which we cannot usefully question. These people are all anxious that we shall behave well, and yet that we shall not question how we behave. Because they believe that there is no rational foundation for values, they fear that an appeal to logic can lead only first to irreverence and then to hedonism.

I do not share this fear, and I do not need it to sustain my sense of values. To me, such a concept as duty is like the concept of mass. I was not born with a concept of mass, and I did not receive it by edict; yet both my inborn senses and my education took part in the process of elucidating it as it grew out of my experience and that of others. I do not find it difficult to defend my concept of mass on these grounds, and I see no reason why I should base the concept of duty as a value on different grounds.

There is I think another fear that moves people to resist the suggestion that the values by which they live should be studied empirically. They grant that this study may indeed reveal what men do in order to prosper; but is this, they ask, what they ought to do? Is it not more often precisely what they ought not to do? Surely, say the righteous, it is the wicked who flourish, and they flourish because they practice what is wicked. So that if social science studies, as natural science does, what works and what does not, the laws which it traces are likely, they fear, to be very unsavory.

I doubt whether this dark view will bear the light of history. Is it really true that the wicked prosper? In the convulsions of nations, have tyrannies outlived their meeker rivals? Rome has not survived the Christian martyrs. Machiavelli in *The Prince* was impressed by the triumphs of the Borgias, and he has impressed us; but were they in fact either successful or enviable? Was the fate of Hitler and Mussolini better? And even in the short perspective of our own street, do we really find that the cheats have the best of it? Or are we merely yielding to the comforting belief that, because one of our neighbors flourishes, he is *ipso facto* wicked?

There is a grave error in this fear that the study of society must reveal a moral form of Gresham's law, that the bad drives out the good. The error is to suppose that the norms of conduct in a society might remain fixed while the conduct of its members changes. This is not so. A society cannot remain lawful when many members break the laws. In an orderly society an impostor now and again gains an advantage; but he gains it only so long as imposture remains occasional—so long, that is, as his own practice does not destroy the social order. The counterfeiter can exploit the confidence of society in the value of money only so long as he himself does not sap this confidence by swamping the market with counterfeits: only so long, that is, as good money remains the norm. Destroy this, and Gresham's law really takes its revenge; the society falls apart to suspicion and barter.

If we are to study conduct, we must follow it in both directions: into the duties of men, which alone hold a society together, and also into the freedom to act personally which the society must still allow its men. The problem of values arises only when men try to fit together their need to be social animals with their need to be free men. There is no problem, and there are no values, until men want to do both. If an anarchist wants only freedom, whatever the cost, he will prefer the jungle of man at war with man. And if a tyrant wants only social order, he will create the totalitarian state. He will single out those who question or dissent— those whom Plato in the *Republic* called poets and in the *Laws* called materialists, and whom Congressional Committees more simply call scientists; and he will have them, as Plato advised, exiled, or *gleichgeschaltet* or liquidated or investigated.

3

The concepts of value are profound and difficult exactly because they do two things at once: they join men into societies, and yet they preserve for them a freedom which makes them single men. A philosophy which does not acknowledge both needs cannot evolve values, and indeed cannot allow them. This is true of a wholly social philosophy such as dialectical materialism, in which the community lays down how the individual must act; there is no room for him to ask himself how he *ought* to act.[3] And it is equally true of the individualist systems which have for some time had a following in England—systems such as logical positivism and its modern derivative, analytical philosophy.

It is relevant to examine these last philosophies, because they make a special claim to be scientific. In their reaction against the metaphysics of the nineteenth century they have returned to the empiricist tradition which goes back in

British philosophy to Thomas Hobbes, to John Locke, and above all to David Hume. This is a tradition which looks for the material and the tests of a philosophy in the physical world; the evidence which it seeks is, roughly, that which a scientist seeks, and it rejects evidence which would not pass muster in science. Those who led the return to the empiricist tradition, first Bertrand Russell and then Ludwig Wittgenstein, were in fact trained in scientific disciplines.

In his early writing Wittgenstein held that a statement makes sense only if it can be tested in the physical world. In his later writing Wittgenstein came to look for the meaning of a statement in the way in which it can be used: the contexts and the intentions into which it fits. That is, his early view of truth was positivist, and his later view was analytical. Wittgenstein's followers have now enthroned his later analysis of usage into a philosophical method which often seems remote from any universal test, but their aim remains, as it was his, to make our understanding of the world tally with the way in which it works in fact.

Positivists and analysts alike believe that the words *is* and *ought* belong to different worlds, so that sentences which are constructed with *is* usually have a verifiable meaning, but sentences constructed with *ought* never have. This is because Wittgenstein's unit, and Russell's unit, is one man; all British empiricist philosophy is individualist. And it is of course clear that if the only criterion of true and false which a man accepts is that man's, then he has no base for social agreement. The question how a man *ought* to behave is a social question, which always involves several people; and if he accepts no evidence and no judgment except his own, he has no tools with which to frame an answer.

The issue then is, whether verification can be accepted as a principle if it is assumed to be carried out by one man. This is as factual an issue as that which faced physics in 1905. Einstein did not debate in 1905 whether space and time may be absolute, in principle; he asked how physicists in fact measure them. So it is irrelevant (and metaphysical) to debate whether verification can be absolute, in principle; the question is, how do men in fact verify a statement? How do they confirm or deny the assertion, for example, that 'The Crab nebula is the dust of a supernova which exploded in 1054, and it glows because some of it is radioactive carbon which was made in the supernova.'

This is a fairly simple speculation, as science goes. The positivist would break it into still simpler pieces, and would then propose to verify each. But it is an illusion, and a fatal illusion, to think that he could verify them himself. Even in principle he could not verify the historical part of this statement without searching the records of others, and believing them. And in practice he could not verify the rate of expansion of the Crab nebula, and the processes which might cause it to glow, without the help of a sequence of instrument makers and astronomers and nuclear physicists, specialists in this and that, each of whom he must trust and believe. All this knowledge, all our knowledge, has been built up communally: there would be no astrophysics, there would be no history, there would not even be language, if man were a solitary animal.

The fallacy which imprisons the positivist and the analyst is the assumption that he can test what is true and false without consulting anyone but himself. This of course prevents him from making any social judgment. Suppose then that we give up this assumption, and acknowledge that, even in the verification of facts, we need the help of others. What follows?

It follows that we must be able to rely on other people; we must be able to trust their word. That is, it follows that there is a principle which binds society together, because without it the individual would be helpless to tell the true from the false. This principle is truthfulness. If we accept truth as an individual criterion, then we have also to make it the cement to hold society together.

The positivist holds that only those statements have meaning which can in principle be verified, and found to be so or not so. Statements which contain the word *is* can be of this kind; statements which contain the word *ought* cannot. But we now see that, underlying this criterion, there is a social nexus which alone makes verification possible. This nexus is held together by the obligation to tell the truth. Thus it follows that there is a social injunction implied in the positivist and analyst methods. This social axiom is that

We OUGHT *to act in such a way that what* IS *true can be verified to be so.*

4

This is the light by which the working of society is to be examined. And in order to keep the study in a manageable field I will continue to choose a society in which the principle of truth rules. Therefore the society which I will examine is that formed by scientists themselves: it is the body of scientists.

It may seem strange to call this a society, and yet it is an obvious choice; for having said so much about the workings of science, I should be shirking all our unspoken questions if I did not ask how scientists work together.[4] The dizzy progress of science, theoretical and practical, has depended on the existence of a fellowship of scientists which is free, uninhibited and communicative. It is not an upstart society; for it derives its traditions, both of scholarship and of service, from roots which reach through the Renaissance into the monastic communities and the first universities. The men and women who practice the sciences make a company of scholars which has been more lasting than any modern state, yet which has changed and evolved as no Church has. What power holds them together?

In an obvious sense, theirs is the power of virtue. By the worldly standards of public life, all scholars in their work are of course oddly virtuous. They do not make wild claims, they do not cheat, they do not try to persuade at any cost, they appeal neither to prejudice nor to authority, they are often frank about their ignorance, their disputes are fairly decorous, they do not confuse what is being argued with race, politics, sex or age, they listen patiently to the young and to the old who both know everything. These are the general virtues of scholarship, and they are peculiarly the virtues of science. Individually, scientists no doubt have human weaknesses. Several of them may have mistresses or read Karl Marx; some of them may even be homosexuals and read Plato. But in a world in which state and dogma seem always either to threaten or to cajole, the body of scientists is trained to avoid and organized to resist every form of persuasion but the fact. A scientist who breaks this rule, as Lysenko has done, is ignored. A scientist who finds that the rule has been broken in his laboratory, as Kammerer found, kills himself.

I have already implied that I do not trace these virtues to any personal goodness in scientists. A recent study [5] has indeed shown that, as a profession, science attracts men whose temperament is grave, awkward and absorbed. But this is in the main the scholar's temperament, which is shared by historians and literary critics and painters in miniature. For all their private virtues, these other men form today what they formed four hundred years ago, scattered collections of individuals. It is not their temperament which has made of scientists so steadfast and so powerful a society.

Nor is it the profession of science, simply as a profession. Every profession has its solemn codes: the lawyers and the salesmen, the accountants and the musicians and the consult-

ing engineers. When a member of these combinations be-
haves outrageously, he is expelled. But this association is as
circumspect as a licensing board and as formal as a trade
union. It guides and it protects the practitioner, it offers
him models and friends, and it gives a personality to his
work. It can be as far-reaching as the Hippocratic Oath
or the ceremonial of Freemasonry. And yet we have only
to see how much alike are all these codes, how pious and
how general, to know at once that they do not spring
from the pith and sap of the work which they regulate. They
are not thrust up, a sharp green bough, from the ruling
passion of their adherents. It is the other way about: their
codes are a reminder to each profession that the sanctions of
society at large reach into them also. Our civilization has
imposed itself on these professions; but no one claims that it
has imposed itself on science.

5

The values of science derive neither from the virtues of its
members, nor from the finger-wagging codes of conduct by
which every profession reminds itself to be good. They have
grown out of the practice of science, because they are the
inescapable conditions for its practice.

Science is the creation of concepts and their exploration
in the facts. It has no other test of the concept than its empirical truth to fact. Truth is the drive at the center of science;
it must have the habit of truth, not as a dogma but as a
process. Consider then, step by step, what kind of society
scientists have been compelled to form in this single pursuit.
If truth is to be found, not given, and if therefore it is to
be tested in action, what other conditions (and with them,
what other values) grow of themselves from this?

First, of course, comes independence, in observation and
thence in thought. I once told an audience of school-children
that the world would never change if they did not contradict

their elders. I was chagrined to find next morning that this axiom outraged their parents. Yet it is the basis of the scientific method. A man must see, do and think things for himself, in the face of those who are sure that they have already been over all that ground. In science, there is no substitute for independence.

It has been a byproduct of this that, by degrees, men have come to give a value to the new and the bold in all their work. It was not always so. European thought and art before the Renaissance were happy in the faith that there is nothing new under the sun. John Dryden in the seventeenth century, and Jonathan Swift as it turned into the eighteenth, were still fighting Battles of the Books to prove that no modern work could hope to rival the classics. They were not overpowered by argument or example (not even by their own examples), but by the mounting scientific tradition among their friends in the new Royal Society. Today we find it as natural to prize originality in a child's drawing and an arrangement of flowers as in an invention. Science has bred the love of originality as a mark of independence.

Independence, originality, and therefore dissent: these words show the progress, they stamp the character of our civilization as once they did that of Athens in flower. From Luther in 1517 to Spinoza grinding lenses, from Huguenot weavers and Quaker ironmasters to the Puritans founding Harvard, and from Newton's religious heresies to the calculated universe of Eddington, the profound movements of history have been begun by unconforming men. Dissent is the native activity of the scientist, and it has got him into a good deal of trouble in the last ten years. But if that is cut off, what is left will not be a scientist. And I doubt whether it will be a man. For dissent is also native in any society which is still growing. Has there ever been a society which has died of dissent? Several have died of conformity in our lifetime.

Dissent is not itself an end; it is the surface mark of a

deeper value. Dissent is the mark of freedom, as originality is the mark of independence of mind. And as originality and independence are private needs for the existence of a science, so dissent and freedom are its public needs. No one can be a scientist, even in private, if he does not have independence of observation and of thought. But if in addition science is to become effective as a public practice, it must go further; it must protect independence. The safeguards which it must offer are patent: free inquiry, free thought, free speech, tolerance. These values are so familiar to us, yawning our way through political perorations, that they seem self-evident. But they are self-evident, that is, they are logical needs, only where men are committed to explore the truth: in a scientific society. These freedoms of tolerance have never been notable in a dogmatic society, even when the dogma was Christian. They have been granted only when scientific thought flourished once before, in the youth of Greece.

6

I have been developing an ethic for science which derives directly from its own activity. It might have seemed at the outset that this study could lead only to a set of technical rules: to elementary rules for using test tubes, or sophisticated rules for inductive reasoning. But the inquiry turns out quite otherwise. There are, oddly, no technical rules for success in science. There are no rules even for using test tubes which the brilliant experimenter does not flout; and alas, there are no rules at all for making successful general inductions. This is not where the study of scientific practice leads us. Instead, the conditions for the practice of science are found to be of another and an unexpected kind. Independence and originality, dissent and freedom and tolerance: such are the first needs of science; and these are the values which, of itself, it demands and forms.

The society of scientists must be a democracy.[6] It can

keep alive and grow only by a constant tension between dissent and respect; between independence from the views of others, and tolerance for them. The crux of the ethical problem is to fuse these, the private and the public needs. Tolerance alone is not enough; this is why the bland, kindly civilizations of the East, where to contradict is a personal affront, developed no strong science. And independence is not enough either: the sad history of genetics, still torn today by the quarrels of sixty years ago, shows that.[7] Every scientist has to learn the hard lesson, to respect the views of the next man—even when the next man is tactless enough to express them.

Tolerance among scientists cannot be based on indifference, it must be based on respect. Respect as a personal value implies, in any society, the public acknowledgements of justice and of due honor. These are values which to the layman seem most remote from any abstract study. Justice, honor, the respect of man for man: What, he asks, have these human values to do with science? The question is a foolish survival of those nineteenth-century quarrels which always came back to equate ethics with the Book of Genesis. If critics in the past had ever looked practically to see how a science develops, they would not have asked such a question. Science confronts the work of one man with that of another, and grafts each on each; and it cannot survive without justice and honor and respect between man and man. Only by these means can science pursue its steadfast object, to explore truth. If these values did not exist, then the society of scientists would have to invent them to make the practice of science possible. In societies where these values did not exist, science has had to create them.

7

Science is not a mechanism but a human progress, and not a set of findings but the search for them. Those who think that science is ethically neutral confuse the findings of science, which are, with the activity of science, which is not. To the layman who is dominated by the fallacy of the comic strips, that science would all be done best by machines, the distinction is puzzling. But human search and research is a learning by steps of which none is final, and the mistakes of one generation are rungs in the ladder, no less than their correction by the next. This is why the values of science turn out to be recognizably the human values: because scientists must be men, must be fallible, and yet as men must be willing and as a society must be organized to correct their errors. William Blake said that 'to be an Error & to be Cast out is a part of God's design.' It is certainly part of the design of science.

There never was a great scientist who did not make bold guesses, and there never was a bold man whose guesses were not sometimes wild. Newton was wrong, in the setting of his time, to think that light is made up of particles. Faraday was foolish when he looked, in his setting, for a link between electro-magnetism and gravitation. And such is the nature of science, their bad guesses may yet be brilliant by the work of our own day. We do not think any less of the profound concept of General Relativity in Einstein because the details of his formulation at this moment seem doubtful. For in science, as in literature, the style of a great man is the stamp of his mind, and makes even his mistakes a challenge which is part of the march of its subject. Science at last respects the scientist more than his theories; for by its nature it must prize the search above the discovery, and the thinking (and with it the thinker) above the thought. In the society of scientists each man, by the process of exploring for the truth, has earned a dignity more profound than his doctrine. A true society is sustained by the sense of human dignity.

I take this phase from the life of the French naturalist Buffon who, like Galileo, was forced to recant his scientific

findings.[8] Yet he preserved always, says his biographer, something deeper than the fine manners of the court of Louis XV;
he kept *'le sentiment exquis de la dignité humaine.'* His
biographer says that Buffon learned this during his stay in
England, where it was impressed on him by the scientists
he met. Since Buffon seems to have spent at most three
months in England, this claim has been thought extravagant.
But is it? Is history really so inhuman an arithmetic? Buffon
in the short winter of 1738–9 met the grave men of the Royal
Society, heirs to Newton, the last of a great generation. He
found them neither a court nor a rabble, but a community
of scientists seeking the truth together with dignity and
humanity. It was, it is, a discovery to form a man's life.

The sense of human dignity that Buffon showed in his
bearing is the cement of a society of equal men; for it expresses their knowledge that respect for others must be
founded in self-respect. Theory and experiment alike become
meaningless unless the scientist brings to them, and his fellows can assume in him, the respect of a lucid honesty with
himself. The mathematician and philosopher W. K. Clifford
said this forcibly at the end of his short life, nearly a hundred
years ago.

> If I steal money from any person, there may be no harm
> done by the mere transfer of possession; he may not feel the
> loss, or it may even prevent him from using the money badly.
> But I cannot help doing this great wrong towards Man, that I
> make myself dishonest. What hurts society is not that it
> should lose its property, but that it should become a den of
> thieves; for then it must cease to be society. This is why we
> ought not to do evil that good may come; for at any rate this
> great evil has come, that we have done evil and are made
> wicked thereby.

This is the scientist's moral: that there is no distinction between ends and means. Clifford goes on to put this in terms
of the scientist's practice:

In like manner, if I let myself believe anything on insufficient evidence, there may be no great harm done by the mere belief; it may be true after all, or I may never have occasion to exhibit it in outward acts. But I cannot help doing this great wrong towards Man, that I make myself credulous. The danger to society is not merely that it should believe wrong things, though that is great enough; but that it should become credulous.

And the passion in Clifford's tone shows that to him the word credulous had the same emotional force as 'a den of thieves.'

The fulcrum of Clifford's ethic here, and mine, is the phrase 'it may be true after all.' Others may allow this to justify their conduct; the practice of science wholly rejects it. It does not admit that the word true can have this meaning. The test of truth is the known factual evidence, and no glib expediency nor reason of state can justify the smallest self-deception in that. Our work is of a piece, in the large and in detail; so that if we silence one scruple about our means, we infect ourselves and our ends together.

The scientist derives this ethic from his method, and every creative worker reaches it for himself. This is how Blake reached it from his practice as a poet and a painter.

He who would do good to another must do it in Minute Particulars:
General Good is the plea of the scoundrel, hypocrite & flatterer,
For Art & Science cannot exist but in minutely organized Particulars.

The Minute Particulars of art and the fine-structure of science alike make the grain of conscience.

8

Usually when scientists claim that their work has liberated men, they do so on more practical grounds. In these four hundred years, they say, we have mastered sea and sky, we have drawn information from the electron and power from the nucleus, we have doubled the span of life and halved the working day, and the leisure we created we have enriched with universal education and high-fidelity recordings and electric light and the lipstick. We have carried out the tasks which men set for us because they were most urgent. To a world population at least five times larger than in Kepler's day, there begins to be offered a life above the animal, a sense of personality, and a potential of human fulfillment, which make both the glory and the explosive problem of our age.

These claims are not confined to food and bodily comfort. Their larger force is that the physical benefits of science have opened a door, and will give all men the chance to use mind and spirit. The technical man here neatly takes his model from evolution, in which the enlargement of the human brain followed the development of the hand.[9]

I take a different view of science as a method; to me, it enters the human spirit more directly. Therefore I have studied quite another achievement: that of making a human society work. As a set of discoveries and devices, science has mastered nature; but it has been able to do so only because its values, which derive from its method, have formed those who practice it into a living, stable and incorruptible society. Here is a community where everyone has been free to enter, to speak his mind, to be heard and contradicted; and it has outlasted the empires of Louis XIV and the Kaiser. Napoleon was angry when the Institute he had founded awarded his first scientific prize to Humphry Davy, for this was in 1807, when France was at war with England. Science survived then and since because it is less brittle than the rage of tyrants.

This is a stability which no dogmatic society can have. There is today almost no scientific theory which was held when, say, the Industrial Revolution began about 1760. Most often today's theories flatly contradict those of 1760; many contradict those of 1900. In cosmology, in quantum mechanics, in genetics, in the social sciences, who now holds the beliefs that seemed firm sixty years ago? Yet the society of scientists has survived these changes without a revolution, and honors the men whose beliefs it no longer shares. No one has recanted abjectly at a trial before his colleagues. The whole structure of science has been changed, and no one has been either disgraced or deposed. Through all the changes of science, the society of scientists is flexible and single-minded together, and evolves and rights itself. In the language of science, it is a stable society.

9

The society of scientists is simple because it has a directing purpose: to explore the truth. Nevertheless, it has to solve the problem of every society, which is to find a compromise between man and men. It must encourage the single scientist to be independent, and the body of scientists to be tolerant. From these basic conditions, which form the prime values, there follows step by step a range of values: dissent, freedom of thought and speech, justice, honor, human dignity and self-respect.

Our values since the Renaissance have evolved by just such steps. There are of course casuists who, when they are not busy belittling these values, derive them from the Middle Ages. But that servile and bloody world upheld neither independence nor tolerance, and it is from these, as I have shown, that the human values are rationally derived. Those who crusade against the rational, and receive their values by mystic inspiration, have no claim to these values of the mind. I cannot put this better than in the words of Albert Schweitzer in which he, a religious man, protests that mysticism in religion is not enough.

Rationalism is more than a movement of thought which realized itself at the end of the eighteenth and the beginning of the nineteenth centuries. It is a necessary phenomenon in all normal spiritual life. All real progress in the world is in the last analysis produced by rationalism. The principle, which was then established, of basing our views of the universe on thought and thought alone is valid for all time.[10]

So proud men have thought, in all walks of life, since Giordano Bruno was burned alive for his cosmology on the Campo de' Fiori in 1600. They have gone about their work simply enough. The scientists among them did not set out to be moralists or revolutionaries. William Harvey and Huygens, Euler and Avogadro, Darwin and Willard Gibbs and Marie Curie, Planck and Pavlov, practised their crafts modestly and steadfastly. Yet the values they seldom spoke of shone out of their work and entered their ages, and slowly re-made the minds of men. Slavery ceased to be a matter of course. The princelings of Europe fled from the gaming table. The empires of the Bourbons and the Hapsburgs crumbled. Men asked for the rights of man and for government by consent. By the beginning of the nineteenth century, Napoleon did not find a scientist to elevate tyranny into a system; that was done by the philosopher Hegel. Hegel had written his university dissertation to prove philosophically that there could be no more than the seven planets he knew. It was unfortunate, and characteristic, that even as he wrote, on 1 January 1801, a working astronomer observed the eighth planet Ceres.[11]

10

I began this book with the question which has haunted me, as a scientist, since I heard it in the ruins of Nagasaki: 'Is You Is Or Is You Ain't Ma Baby?' Has science fastened upon our society a monstrous gift of destruction which we can neither undo nor master, and which, like a clockwork automaton in a nightmare, is set to break our necks? Is science an automaton, and has it lamed our sense of values?

These questions are not answered by holding a Sunday symposium of moralists. They are not even answered by the painstaking neutralism of the textbooks on scientific method. We must indeed begin from a study of what scientists do, when they are neither posed for photographs on the steps of spaceships nor bumbling professorially in the cartoons. But we must get to the heart of what they do. We must lay bare the conditions which make it possible for them to work at all.

When we do so we find, leaf by leaf, the organic values which I have been unfolding. And we find that they are not at odds with the values by which alone mankind can survive. On the contrary, like the other creative activities which grew from the Renaissance, science has humanized our values. Men have asked for freedom, justice and respect precisely as the scientific spirit has spread among them. The dilemma of today is not that the human values cannot control a mechanical science. It is the other way about: the scientific spirit is more human than the machinery of governments. We have not let either the tolerance or the empiricism of science enter the parochial rules by which we still try to prescribe the behavior of nations. Our conduct as states clings to a code of self-interest which science, like humanity, has long left behind.

The body of technical science burdens and threatens us because we are trying to employ the body without the spirit; we are trying to buy the corpse of science. We are hagridden by the power of nature which we should command, because we think its command needs less devotion and understanding than its discovery. And because we know how gunpowder works, we sigh for the days before atomic bombs. But massacre is not prevented by sticking to gunpowder; the Thirty

Years' War is proof of that. Massacre is prevented by the scientist's ethic, and the poet's, and every creator's: that the end for which we work exists and is judged only by the means which we use to reach it. This is the human sum of the values of science. It is the basis of a society which scrupulously seeks knowledge to match and govern its power. But it is not the scientist who can govern society; his duty is to teach it the implications and the values in his work. Sir Thomas More said this in 1516, that the single-minded man must not govern but teach; and nearly twenty years later went to the scaffold for neglecting his own counsel.

11

I have analyzed in this book only the activity of science. Yet I do not distinguish it from other imaginative activities; they are as much parts one of another as are the Renaissance and the Scientific Revolution. The sense of wonder in nature, of freedom within her boundaries, and of unity with her in knowledge, is shared by the painter, the poet and the mountaineer.[12] Their values, I have no doubt, express concepts as profound as those of science, and could serve as well to make a society—as they did in Florence, and in Elizabethan London, and among the famous doctors of Edinburgh. Every cast of mind has its creative activity, which explores the likenesses appropriate to it and derives the values by which it must live.

The exploration of the artist is no less truthful and strenuous than that of the scientist. If science seems to carry conviction and recognition more immediately, this is because here the critics are also those who work at the matter. There is not, as in the arts, a gap between the functions (and therefore between the fashions) of those who comment and those who do. Nevertheless, the great artist works as devotedly to uncover the implications of his vision as does the great scientist. They grow, they haunt his thought, and their most inspired flash is the end of a lifetime of silent exploration. Turn to the three versions of *Faust* at which Goethe worked year in and year out. Or watch Shakespeare at work. Early in this book I quoted from *Romeo and Juliet* the image of death as a bee that stings other people, but that comes to Juliet to drink her sweetness—

Death that hath suckt the honey of thy breath.

More than ten years later Shakespeare came back to the image and unexpectedly made it concrete, a metaphor turned into a person in the drama. The drama is *Antony and Cleopatra;* the scene is the high tower; and to it death comes in person, as an asp hidden among figs. The image of the asp carries, of course, many undertones; and most moving among these is Cleopatra's fancy, that this death, which should sting, has come to her to suck the sweetness. Cleopatra is speaking, bitterly, tenderly, about the asp:

Peace, peace:
Dost thou not see my Baby at my breast,
That suckes the Nurse asleepe.

The man who wrote these words still carried in his ear the echo from Juliet's tomb, and what he added to it was the span of his life's work.

Whether our work is art or science or the daily work of society, it is only the form in which we explore our experience which is different; the need to explore remains the same. This is why, at bottom, the society of scientists is more important than their discoveries. What science has to teach us here is not its techniques but its spirit: the irresistible need to explore. Perhaps the techniques of science may be practiced for a time without its spirit, in secret establishments, as the Egyptians practiced their priestcraft. But the inspiration of science for four hundred years has been opposite to this. It has created the values of our intellectual life and, with the arts, has taught them to our civilization. Science has nothing to be ashamed of even in the ruins of Nagasaki. The shame is theirs who appeal to other values than the human imaginative values which science has evolved. The shame is ours if we do not make science part of our world, intellectually as much as physically, so that we may at last hold these halves of the world together by the same values. For this is the lesson of science, that the concept is more profound than its laws, and the act of judging more critical than the judgment. In a book that I wrote about poetry I said,

> Poetry does not move us to be just or unjust, in itself. It moves us to thoughts in whose light justice and injustice are seen in fearful sharpness of outline.[13]

What is true of poetry is true of all creative thought. And what I said then of one value is true of all human values. The values by which we are to survive are not rules for just and unjust conduct, but are those deeper illuminations in whose light justice and injustice, good and evil, means and ends are seen in fearful sharpness of outline.

NOTES

1. This view, that science is as integral to our culture as the arts, and as necessary to our education, was the theme of my address to the British Association for the Advancement of Science in 1955 'The Educated Man in 1984' and of Sir Charles Snow's eloquent Rede Lecture *The Two Cultures and the Scientific Revolution* (Cambridge, 1959).

2. It is a common and cardinal error to suppose, as the nineteenth century supposed, that the facts on which science builds are given to us absolutely,

and call for no judgment or interpretation from us. The great discoveries in the physical sciences in the twentieth century begin from a radical denial of this philosophy. We now understand that science is built not on facts but on observations; that observation is not a passive state of reception, but an active relation between the observer and his world; and that science is therefore not a mechanical index of facts, but an evolving activity. These principles have recently been presented in detail by Michael Polanyi in *Personal Knowledge* (Harper Torchbooks, 1964).

3. This outlook is evident in the genuine distress which Boris Pasternak's novel *Dr. Zhivago* caused to good literary critics in Russia. Their official letter of rejection is temperately and indeed reasonably argued, once we accept its overriding assumption that Dr. Zhivago himself has no business to continue to hold to his personal values (to his poems for example) when the lives of his fellow citizens are being torn by change. Thus what is being judged is not *Dr. Zhivago* as a book, but Dr. Zhivago as a character who cannot conform.

4. Because scientists have constantly to use one another's work, and to rely on one another's good faith, they have a strong sense of belonging to a community. They are very conscious of their isolation if they reject the code of conduct of the scientific society. Thus Klaus Fuchs in his confession finds it worthy of remark that when he had disclosed the secrets of Los Alamos to the Russian agent 'Raymond' in 1945, 'It appeared to me at the time that I had become a "free man" because I had managed to establish myself in an area of my being as completely independent of the surrounding forces of society.' Fuchs is here picturing his disclosures almost as if they had been an *acte gratuit*, whose purpose was to demonstrate by arbitrary means his rejection of all social ties. This state of mind is remarkable in Fuchs as a scientist; it would not be extraordinary in an existentialist, committed only to the uniqueness of his own life. Fuchs's confession invites comparison with the self-searchings of some characters in Jean-Paul Sartre's play *Les Mains Sales*, and indeed many of the same loyalties are engaged in that play.

5. I have in mind the work of Anne Roe, as described in her book *The Making of a Scientist* (New York, 1953) and elsewhere.

6. This limits the actions which the individual scientist thinks himself entitled to impose on others. He does not think himself entitled to play the benevolent despot, who can insist that he knows better than anyone what is good for everyone. By contrast, many non-scientists believe that the misuse of science can only be ended if individual scientists take away the responsibility for its proper use from the community, and decide to withhold from it the knowledge of discoveries which might be misused. On this view Einstein in 1939 should not have told the President of the United States that an atomic explosion might be produced—on the ground, presumably, that

Einstein could be trusted to act with wisdom and humanity, and the man whom the United States had elected to act for them could not. In the event, however, the only scientist who aspired to dictate to the communal conscience in this way turned out to be Klaus Fuchs. I have discussed this issue at length in an essay on 'Responsibilities of Scientist and Public' in the *Atomic Scientists' Journal* for September 1955.

7. The feud that was waged by Lysenko against the majority of geneticists in Russia in recent years had a parallel in a bitter quarrel that divided English geneticists about 1900. At that time, however, the roles were reversed: most biologists believed that hereditary variation is continuous, and it was the rebels who believed in the new Mendelian mechanism by which heritable characters are passed on as units. The rebels were led by William Bateson, who was (like Lysenko) a naturalist and a breeder, a man with green fingers. Like Lysenko, Bateson argued against what seemed to him the lifeless statistics of laboratory genetics; but the statistical studies he complained of were of continuous variations, and precisely opposite to those now attacked by Lysenko. Among those ranged against Bateson then was Karl Pearson, a man as great and original, and as intransigent, who would not print Bateson's papers in *Biometrika*: for geneticists and statisticians seem always to have been (and remain) difficult and intemperate men. Yet at bottom (and this is the irony in their quarrel) Pearson's exact statistical methods were then as radically new, and as distasteful to academic biologists, as was Bateson's advocacy of Mendel—and indeed, Pearson and his friends had founded *Biometrika* because the Royal Society made difficulties about publishing their work. There was a fierce showdown between the two schools at the British Association in 1904, much like the showdown at the Academy of Agricultural Sciences in Russia in 1948, with this difference: that at the British Association no one was humiliated, no one was silenced, and no one tried to change genetics on any other ground than the experimental evidence.

8. These findings concerned, among other things, the origin of the earth. Buffon speculated intelligently on a number of related topics in biology and geology, and was probably the first scientist who clearly conceived a theory of evolution.

9. This is the conclusion drawn from the fossil finds of Raymond Dart, Robert Broom and L. S. B. Leakey in South Africa, where it now appears likely that modern man originated.

10. I quote from *The Philosophy of Civilization—Part I, Decay and Restoration* (London, 1932). I have discussed modern rationalism in my Conway Memorial Lecture 'The Fulfilment of Man' (1954).

11. The working astronomer was Giuseppe Piazzi. I owe my knowledge of this bizarre incident to the most distinguished pioneer in the study of the history

of science, the late George Sarton. The just comment on such speculations was made long before by Shakespeare:

> *Foole.* The reason why the seuen Starres are no mo then seuen,
>   is a pretty reason.
> *Lear.* Because they are not eight.
> *Foole.* Yes indeed, thou would'st make a good Foole.

12. In science and in the arts the sense of freedom which the creative man feels in his work derives from what I have earlier called the poetic element in it: the uninhibited activity of exploring the medium for its own sake, and discovering as if in play what can be done with it. The word play is in place here, for the play of young animals is of this kind—an undirected adventure in which they nose into and fill out their own abilities, free from the later compulsions of need and environment. Man plays and learns for a longer time (he has a longer childhood) and he goes on playing into adult life: in this sense of free discovery, pure science is (like art) a form of play. I have analyzed the sense of freedom in another field which lies on the boundary between science and art, in an essay entitled 'Architecture as a Science and Architecture as an Art' which appeared in the *Journal of the Royal Institute of British Architects* of March 1955.

13. *The Poet's Defence* (Cambridge, 1939).

The Abacus
and the Rose:

A New Dialogue on Two World Systems

The characters in the Dialogue are

SIR EDWARD ST. ABLISH, who represents the Establishment: urbane and maddeningly tolerant, fifty-five-plus, Deputy Secretary to the Ministry of Education (or the like), acquired Oxford voice with Edinburgh base.

DR. AMOS HARPING, who represents the literary furies, something between *Scrutiny* and Jimmy Porter: a puritan anger, but bitter because he feels helpless in a changing time—about thirty-five, Reader in English at Southampton, say, Midland voice of preacher with cutting edge.

PROF. LIONEL POTTS, FRS, who represents science: a little smug because success came young, and slow to see that there really are other points of view (and interests) than that of the molecular biologist—not yet forty-five, lacking the critical gift of the other two, his sense of mission as sharply positive as theirs is negative, Irish voice smoldering with idealism.

The three characters are attending an East-West conference on some cultural topic at Lucerne in Switzerland; their expenses have been paid by Her Majesty's Government (perhaps reluctantly) at the request of Unesco. Sir Edward has been troubled by the discord between his two chief delegates (he feels that Britain should speak with one voice) and he is taking them out to dinner to ask them to be, or at least to appear, friendly. He has chosen the restaurant at the top of the Rigi, and they have just arrived there on the rack railway. The restaurant is popular and not at all select, and the noises from time to time recall this: waiters, the laboring rack-and-pinion, picnickers inclined to yodel, motorcoaches hooting to collect their scattered passengers.

SIR EDWARD: I telephoned for this table because I remembered the sunset. Professor Potts, you must sit here, on my right, where the view is best. . . . Will you take the other side, Dr. Harping? And turn your chair around a little, do; you must see the sunset. That is what I brought you for, both of you.

HARPING: Sir Edward, you said that all the way up in the train. The view is fine here; do I have to crane my neck at it all the time? Why do people with knighthoods always behave as if they had personally arranged the sunset?

SIR EDWARD: Come, Dr. Harping, that is not worthy of your wit. I did not arrange for the sun to set. But I did personally arrange for you to be here when it did set.

POTTS: Bravo, Sir Edward. And that, my dear Harping, is called administration.

SIR EDWARD: Indeed it is, Professor Potts: administration. And I am proud to put it to the service of the sunset. Isn't it beautiful? That glow of flamingo red on the snow, and then the hard scarlet flash behind the peaks.

HARPING: Huh!

SIR EDWARD: Do back me up, Potts. These literary men no longer care for beauty.

POTTS: You put me in a false position, Sir Edward. Not appreciating the beauty of the Alps is an old story, you know. Samuel Butler tells it.

SIR EDWARD: Really? I didn't know.

HARPING: Oh Potts, not that Victorian chestnut!

POTTS: A friend of Butler's praised the scenery to a Frenchman who was staying in the same hotel. The Frenchman said politely, "Oh, do you like the beauties of nature? I detest them."

SIR EDWARD: I see I am in for a difficult evening with the two of you. I brought you here to ask you to be nice to each other. I hoped you could find some common ground.

HARPING: We have a common ground, Sir Edward. We like being beastly to you.

SIR EDWARD: I won't listen to you, Harping. I am about to order you a glass of sherry. I shall order the sherry that I like; kindly have the grace to like it too. And do not read me a sermon on the evils of alcohol, either of you.

HARPING: Potts, I want to ask you a question. If the sunset leaves you cold, why don't you say so? Why do you need Samuel Butler to give you courage?

POTTS: Because Samuel Butler is a hero of mine. He was full of fancies and he quarreled with Darwin, but he had an interesting mind.

HARPING: Nothing about his story is interesting. He wasn't even the one who detested nature, remember? It was the Frenchman. You just drag in Samuel Butler because you want literary support. Why? You're supposed to be a scientist. Why this parade of the well-read, well-thumbed literary sources? These are tidbits that my first-year students know, even if Sir Edward doesn't. You don't need to show off that you know them too.

POTTS: Of course not, Harping; I don't show off. I am simply used to quoting my sources. If Samuel Butler made the original observation, then I quote Samuel Butler. And he was a most original observer, you must agree.

HARPING: He was not. He was an eccentric, so you think he was original. That's the judgment of science. And he disliked authority, so you quote him as an authority. That's the scientific attitude. And really, it's so terribly literary—like Latin tags and stories about Dr. Johnson. You know, Samuel Butler had a passion for new-laid eggs. I shall avoid having breakfast with you, Potts, for fear that you will feel it necessary to bore me with his example in order to justify your second egg.

SIR EDWARD: Here's the sherry, gentlemen, and please

drink it temperately. It's a triumph to get sherry before dinner in Switzerland—yes, I know, Professor Potts, a triumph of administration.

POTTS: A triumph that I appreciate, Sir Edward.

*

SIR EDWARD: I am sure you were glad of the diversion, Potts. All the same, Harping, I don't think you were fair. Why be cruel to Potts's heroes? You have your heroes too, I remember; I have heard you being very tedious about D. H. Lawrence—and at breakfast, too. Is there anything to choose between them? Butler the shy eccentric of the head, and Lawrence the stern eccentric of the heart. Admit it, Dr. Harping: the heart has its unreasons too!

HARPING: Sir Edward St. Ablish, are you seriously suggesting that the profound and professional craft of D. H. Lawrence is to be compared with the dilettante talents of—

SIR EDWARD: Please don't get up, Dr. Harping; there is no train back for some time. I am suggesting that *The Way of All Flesh* is a novel of the same excellence as *The White Peacock*, certainly. But you miss my point if you turn it into an argument about literary merit.

POTTS: And what is your point, Sir Edward?

SIR EDWARD: My point, Professor Potts, is that you both need heroes, Dr. Harping and you. Harping has been scolding you because you cannot be unconventional on your own two feet. You have a solemn convention for being unconventional—it's the scientific convention. But Harping has his own convention too for being unconventional. He takes his tie off when he lectures; and I dare say *you* have some trick of stripping off your white coat in times of drama in your laboratory. You pioneers have to be so individual, both of you. Harping has to look like his teacher—what's his name?—Dr. Leavis; and you have to look like an Irish groom.

HARPING: I don't deny that, Sir Edward, and it's not very deep. We all have our patterns. Of course we do. Potts has his kind of nonconformity, and I have mine. And you have your uniform of cozy, tolerant conformity that mocks at heroes. But I wasn't taxing Potts with following a pattern; I was criticizing the particular pattern that he follows. It's such a literary and derivative pattern. Why call yourself a scientist and then fill your mind with a ragbag of nineteenth-century sentiments? Is that supposed to be a union of two cultures? You don't know what literature is about, Potts. Culture, indeed; all you collect is a string of cultured pearls.

POTTS: I really don't see your evidence for that, Harping. Do you always get carried away by one example? We were asked, both of us, if we thought the sunset beautiful. You don't like my answer. Very well, let me hear yours. I notice that you have avoided saying anything yourself. Do you find the sunset beautiful?

*

HARPING: You can't be serious. You don't really expect me to talk about beauty in sunsets, do you?

POTTS: Why not?

SIR EDWARD: Why not, indeed, Dr. Harping? I asked the question, and certainly I expect a serious answer. Here it is, the last of the sunset glowing over the Alps. Rose-red of snow on crimson, and vermilion over gold. In ten minutes it will be gone; and in half an hour it will be dark. It's a symbol of our conference: the West being drenched with red and after that, darkness. It deserves your attention, Dr. Harping, surely; I wish I could get you to give it your joint attention, both of you. So, in all seriousness, I should like to have your reflections on the sunset, Dr. Harping.

HARPING: But how can you ask such a question? It doesn't make sense. Is the sunset beautiful? What a word to use! The sunset isn't an object like a coin, and beauty isn't a value that has been stamped on it, officially: one sunset, mint condition, like a half-crown.

POTTS: You're always protesting, Harping. Here you are, full of scorn, telling us what the sunset is not. All right, I know what the sunset is not. It's not a half-crown, and I never thought it was. And I did not ask you what the sunset is not. Just tell me what it is. Tell me if it's beautiful, and why.

HARPING: But the question doesn't mean anything, Potts. I would rather not use words like beauty; but if that's the word you want to use, very well. Beauty is not a currency; it can't be passed from hand to hand at an agreed value; and it certainly is not absolute by nature. Beauty is a personal relation; something that the sunset says to me, or doesn't say; and that I can then discuss with you, and shed light on for you, and you for me. But I cannot hand it over to you as a finished judgment. I can hand you a copy of *The White Peacock,* Sir Edward, but I cannot hand you its greatness. You can arrange for me to see the sunset, but you cannot arrange for me to see its beauty.

SIR EDWARD: I understand that, Dr. Harping, and if I may borrow your phrase, it's not very deep. You have something deeper to explain to me: why what the sunset says to you and what it says to me cannot be contained in a common description. Beauty is not absolute by nature, you said — I think you said 'certainly,' as people do when they want to avoid discussion. Did you say 'certainly'?

HARPING: I have no doubt I did.

POTTS: He would prefer you to accept the proposition as self-evident rather than as reasonable.

SIR EDWARD: I do not think that beauty is absolute either, Dr. Harping; I have learned that it is not, by experience. But I should like to have the proposition discussed. It would make an interesting session of the conference: more interesting than all the talk about culture and democracy. How we chew those two words, culture and democracy, like two fried eggs—do you prefer them Western style, sunnyside up, or simply turned over, Eastern style? That reminds me, I have ordered you an *hors d'oeuvre*. It's very Swiss, I'm afraid; it has cheese in it. Waiter! We will have the *hors d'oeuvre* now, and another glass of sherry. Can you bear to agree with the absolute judgment of all wine tasters, Dr. Harping, that this is a beautiful sherry?

POTTS: Harping will tell you that you do not mean the absolute judgment, Sir Edward: you mean the universal judgment. And you do not mean a beautiful sherry, either; you mean a splendid sherry.

HARPING: All right, Potts, tease as much as you like: it still is so. There are things you can say about the sunset which are plain enough, crude enough, to be universal. The sunset is splendid, meaning that (like the sherry) it is full of gold. The sunset is spectacular, meaning that it is more color-ful than our daily view of the sky. You can even say that the sunset is exhilarating because it gives all of us a sense of the grandeur of nature which we lack at other times. These are statements which are rough, direct and, at bottom, de-scriptive. They are essentially statements of fact, which com-pare the brightness before us with the more ordinary condi-tions of life that we all share.

SIR EDWARD: And yet I must not say that the sunset is beau-tiful?

HARPING: No, Sir Edward, you must not; you must not, if you want to use words sensitively. Beauty is not measured like splendor, by a comparison with the commonplace. It is felt in each of us by what is most individual in him. When we discuss beauty as we should, we are not looking for com-mon ground, the way you are, Sir Edward, by profession—and the way Potts is always looking for common ground, too, because that's all that scientists understand. My profession is to discuss beauty—if that's the word that we've decided to embarrass ourselves with this evening. And when we in my profession, the critics, when we discuss beauty in a work of art or of nature, we are looking for what is uncommon, what is personal, what we can see and someone else cannot. We are using the occasion, each of us, to search into his own individual gifts of appreciation.

SIR EDWARD: Why? You may be engrossed in your private exercise; but why should I be? Why should I attend to what you, Amos Harping—you and no one else—have to say about a poem or a sunset?

POTTS: He may be a better judge of them than you are, Sir Edward.

HARPING: No, Potts, no, no—or rather, yes and no. Yes I am a better judge because no I am not a better judge—because I am aware that no one is a better judge. Why did you ask me about the sunset, Sir Edward? Do you think me a better judge of sunsets than you?

SIR EDWARD: No, Harping, I do not.

HARPING: Then why did you ask me?

SIR EDWARD: Because I wanted to hear what you would say.

HARPING: Of course; and that is the nature of critical appreciation. You want to hear what I say, not because it is better than what you say, but because it is different—minutely, subtly different, different in this personal foible or in that glimpse of another mind. And these differences, these small flashes of light behind the outline, they illuminate and enrich your own vision. You wanted to hear what I had to say, Sir Edward, because now that you have heard it, you will make it your own—a shift of emphasis, an infinitesimal enlargement of your apparently sacred and settled opinion of sunsets.

*

SIR EDWARD: Elegantly put, Dr. Harping. I congratulate you; you can be persuasive when you choose. I wish you would choose to be persuasive in the conference hall; I need your help, both of you, as a matter of British prestige. You know that. Professor Potts, tell me, has Harping persuaded you?

POTTS: He has not persuaded me.

SIR EDWARD: And why not?

POTTS: Because everything he says is obvious, and he makes it seem so elaborate. He says so much, and then it really doesn't mean much. It's all right to call the sunset splendid, but it's not right to call it beautiful. Is that supposed to persuade me? Why?

HARPING: Surely I've explained that. 'Splendid' is a description, but 'beautiful' is a judgment.

POTTS: And am I supposed to be persuaded by those names? Description versus judgment: do you really ask me to accept that shallow distinction?

HARPING: Oh no, oh no, I do not ask you to accept it. Critics know the difference between a description and a judgment, and philosophers know it. Sir Edward St. Ablish knows it. But not you, not Professor Lionel Potts, FRS. Because Lionel Potts is a scientist. He does not accept fine distinctions. Lionel Potts sits in his laboratory and peers down his microscope and describes what he sees; and everything that Lionel Potts sees can be described. Does Lionel Potts see the sunset? It can be described. It is splendid—of course it is, because the word means 'shining.' A good, solid, half-crown word; you can weigh the sunset with it to the nearest gram. That's the way Lionel Potts sees things in the Royal Society. It's a sunset; it's splendid; it's red. What wavelength, Professor Potts?

POTTS: Around 7,000 angstroms.

HARPING: The answer of a scholar: around 7,000 angstroms. And the sun is ninety thousand miles away—

POTTS: Ninety million miles—

HARPING: My mistake. I apologize for my crass mistake.

I'm sure it matters to someone. The sun is splendid, it's setting on a wavelength of 7,000 angstroms, which is red; it is ninety million miles away, and it weighs—what does the sun weigh, Professor Potts?

POTTS: I don't remember.

HARPING: Never mind, I won't tell the Royal Society. Professor Lionel Potts doesn't know what the sun weighs, but he knows it weighs something. Something exact, to three places of decimals. Lionel Potts knows that everything weighs something. Everything can be measured and photographed and spectrographed and God-knows-what-o-graphed. That's it: everything in Lionel Potts's world can be graphed—just graphed. Everything can be described. Who would dare to tell Professor Lionel Potts, FRS, that beauty cannot be described? Who could hope to persuade him that description is not enough; that life, life outside the laboratory, also calls for judgments? . . .

POTTS: Are you pausing for an answer, Harping? Or are you only pausing for breath?

SIR EDWARD: He is pausing for effect, Potts. But I want to hear your answer.

HARPING: Why trouble him, Sir Edward? Potts is a scientist; why trouble us with his thoughts? He is a man of action —the twentieth-century man of action. He doesn't know what a judgment is.

POTTS: On the contrary, Harping, I don't know what a description is.

*

SIR EDWARD: What's that, Potts? What did you say?

POTTS: I said, on the contrary, I don't know what a description is. I know what a judgment is; at least, I know when I make a judgment. But I don't know how to make a description—a scientific description, a pure, precise, mechanical description; a description, period, with no judgment in it.

SIR EDWARD: Potts—I would say solemnly Professor Lionel Potts, except that Harping has charged the name with such extraordinary passion—Professor Potts, I do not know what you mean, and I should like you to tell me.

Please stop for a moment. We are to have trout and Pouilly Fumé. The service is indifferent, but that is the penalty that we pay for being on top of a mountain. I have observed (is that the neutral word I ought to use?)—I have observed that in Switzerland the quality of the service varies inversely with the altitude. Is that a description in your philosophy, Harping; or is it a judgment? No, I mustn't be trivial; I apologize to you both. Potts, you are going to explain why you cannot separate a description from a judgment.

POTTS: Gladly, Sir Edward, and very easily. Harping says that I look down a microscope and describe what I see. I say that I describe what I judge that I see.

SIR EDWARD: Is that an important difference?

HARPING: Of course it is not, Sir Edward. Potts is simply using a trick of speech. What he describes is what he sees or what he believes that he sees; it has nothing to do with judgment.

POTTS: It has everything to do with judgment. Harping tells me that the sunset is red and then asks me contemptuously for the wavelength. But I am not a machine—I *use* a machine. That is, I look at the record and then I judge what I shall say. Shall I give Amos Harping the wavelength to a fraction of an angstrom? And if so, what fraction? Or will it do to give it to him (and to myself) to the nearest hundred angstroms? In fact, I gave it to the nearest thousand. And that in itself was a judgment, Harping—it was my judgment of your seriousness.

HARPING: Potts, you astonish me. Am I to take you seriously? I talk about profound and human judgments of nature and of art. And you tell me, with a straight face, that reading a slide rule is an act of judgment too, because you have to make up your mind, your immense and cosmically learned mind, what is the nearest whole number that you will quote from the slide rule. How trivial—no, I beg your pardon, Sir Edward; you used that word. How silly, how fundamentally silly can you be, Potts?

POTTS: Not how silly, Harping: the question is, how fundamental can I make you be. You think that it is trivial to decide whether I shall read a slide rule to one decimal place or to two. What does it matter, you say; one makes whatever approximation is convenient. You are mistaken—deeply mistaken. Every approximation in science is a judgment; it asserts that the analysis need be pushed no further. We can stop here, it says; what has been left out is not important. And what does 'important' mean? It means that what has been put in is judged to be relevant, and what has been left out is judged not to be relevant. Every scientific description is of this kind: it puts down what it judges to matter, and it leaves out what it judges not to matter. And then one day a scientist with a more adventurous mind makes a new judgment. He decides to push the analysis a step further. He is not content to accept that only so much matters; he takes in another detail. And behold, when the new detail is included, the picture of the universe that accommodates it is wholly different.

SIR EDWARD: Wholly different, Professor Potts? Is it really very different? Does the picture of the universe really change much because a scientist looks at the next decimal point?

POTTS: Yes, Sir Edward, the picture is transformed. Relativity, as an equation, is only minutely different from Newton's physics; but as a picture of the universe, it is fantastically different. What did Einstein do? Very little. He refused to accept the established judgment of others, that time is an irrelevant constant.

HARPING: I don't believe that. I don't believe that Einstein discovered Relativity by arguing about another place of decimals. And if he did: if the approximation of a scientist is a judgment, then I have still to be convinced that it is a searching judgment—a judgment which truly involves the man's personality. My butcher makes an approximation every time he slaps a piece of meat on the scale and tells me the price. Is he a judge, too, because his pocket is involved?

POTTS: A man's adherence to science, his acceptance of that picture of the universe, that is the involvement. Science is an integrated vision, and even when I quote a wavelength, I judge every part of science. I judge that the parts form a unity. There are no facts without that unity, no approximations, and no descriptions. That is why I called your distinction a shallow one, Harping. I assure you, my microscope shows nothing that a machine could turn into a description of life. Science is not made by machines, but by men. And the men in my laboratory are as deeply involved, are as wholly parts of their own judgments, as any student of yours who discusses Shakespeare's *Sonnets* with you.

*

SIR EDWARD: I understand you, Potts. You say that a scientist, or a writer, cannot describe what he sees without editing it, rounding it off, interpreting it; and that every such interpretation is a judgment. The young men in your laboratory, you say, are involved in these judgments, deeply, as human beings. But do you think, Potts, that what the men in your laboratory discover is of the same depth, the same intellectual depth and human complexity, as the *Sonnets*?

POTTS: The men in my laboratory are not all Shakespeares, Sir Edward. And neither are Harping's students. You must compare like with like. The journeymen of science, the journalists of literature, exist in every field. What my young men discover is no doubt about as good as the poetry that Harping's students write.

SIR EDWARD: I am sorry, Potts; I put my question badly. I will try again. Is the vision (I think you called it a vision) that your young men have of nature—is that of the same intellectual depth and human complexity as that which the *Sonnets* present?

POTTS: Of course it is. The vision that science presents of the physical world now, in the second half of the twentieth century, is, in its intellectual depth, its complexity and articulation, the most beautiful and wonderful collective work of the mind of man.

HARPING: A round and noble sentiment. Am I supposed to be awed by the thought of Professor Lionel Potts reverently contemplating, once a day or so, the intellectual depth, complexity and articulation of molecular biology?

SIR EDWARD: Don't be childish, Harping; don't sneer. Do you think it is more awe-inspiring, or less ridiculous, to think of Dr. Amos Harping contemplating, once an hour or so, the emotional depth, complexity and articulation of D. H. Lawrence? We are not schoolboys, Harping; we are not even undergraduates, Potts and I. You can't make us blush for our souls by asking us at what time of day we air them.

HARPING: But I do ask you, Sir Edward; I ask you both. Here is Potts parading his fine phrases about the intellectual depth, the complexity, the articulation of science. Not even his own fine phrases: C. P. Snow's, I think, or else Bronowski's. And you are impressed, and I am expected to be impressed, because he or the phrase-makers that he quotes have shown—have shown what, Sir Edward? They have shown that science is an intricate and neatly jointed construction. Is this supposed to astonish and silence me? Is this supposed to compare with the wealth of Shakespeare and the warmth of Lawrence? Science is an intricate and neatly jointed construction. Contemplate that once a day, or once an hour, or every other minute. What share do you have in it? Yes, you, Potts, or your students, or your fellow scientists? It is an edifice built almost entirely by others, piecemeal, brick by brick, to which you have added nothing but another brick. This is not a vision of nature; it is a description which is presented to you ready-made, and you accept it and look for another brick. Do you seriously compare this, Potts, with the sense of community that my students get when they discuss either Shakespeare or Lawrence? My students (don't you sneer at them, Sir Edward), my students have to make their own discoveries, have to find and share and argue their own meanings. What is on the page is only the beginning of their understanding. They do not simply accept the work of art; they re-create it.

*

POTTS: Amos Harping, you really are the most fervent, the most devoted and the blindest man that I know. Your students are the salt of the earth; do you think mine are less? Your students stand on tiptoe and reach for the light. Do you think mine do less? Your students argue and discover and re-create every tittle on the page that you read with them. Do you think, can you think, that my students do less? My students wrestle with me the way your students wrestle with you—neither more nor less. My students don't read Charles Darwin any less critically than yours read Christopher Marlowe; and they add as much to Darwin's meaning, they enter as avidly into his mind, as yours do. Good God, Harping, you should come to one of my seminars. Why, my students (my good students) won't even believe Mendel's laws without searching their conscience late into the night.

SIR EDWARD: You must explain that to me, Professor Potts. You're called a molecular biologist. I don't quite know what that is, but I imagine that it makes you a biologist of some kind. Well, to me, and I suspect to Harping too, biology is still something of a no man's land. We aren't sure that Darwin and Mendel were altogether scientists, as we think of exact science. So will you give us an example from physics? Something simple and practical that we learned at school. You know, like Ohm's law—the current in a wire and all that. Are you saying that physics students are allowed to argue about the current in a wire?

POTTS: Of course, Sir Edward. Why should an electric current be sacred? My own students argue about Ohm's law like mad—that is, if they're any good. Only a dull student takes Ohm's law on trust and learns it by heart: the current in a wire is proportional to the potential difference, and inversely proportional to the electrical resistance. I have no doubt that Harping's duller students learn bits of Joseph Conrad by heart, too, and make good practical use of them when they become schoolmasters or reviewers. But a bright student who wants to know what a molecule is does not take even Ohm's law on trust.

HARPING: You mean that he goes into the laboratory and tests it for himself with a piece of wire and a battery. That is the extent of his personal participation.

POTTS: No, Harping, it is not. A bright student wants to know why a current acts in this way. He wants to know why, all over the universe, currents flow like this and only like this. What is the nature of a current, he asks, that it has this faithful identity—in my copper wire, in the headlights of your car, in the collision of nebulae, in Sir Edward's gold pin, and in a ring of lead near the absolute zero of cold? The good students put it to me that Ohm's law has to agree, to link and make sense, with what they have learned about electrons, about the structure of metals, and about the picture of the atom that Rutherford created.

HARPING: But what does the student get from this, Potts?

What do you and your fellow scientists get from it? You confirm that the description of nature that has been made by others is internally consistent. Is that a profound experience? Science is an intricate and neatly jointed construction; I know that, I said it. And I ask again: Is the realization of this what a scientist calls a profound experience?

POTTS: There are two questions there, Harping: one about nature and the other about scientists. I will answer the one about nature because to me it is fundamental. When the student sees that Ohm's law flows from the behavior of electrons, when he finds that it is part of, and that it expresses, the structure of matter itself, then he does have a profound experience. He knows that nature is one in all her expressions. And he knows that he is one with nature—he is one of her expressions. It is not the intricacy of nature that moves him then, Harping—what you call the neat jointing, the clockwork mechanics. It is the sweeping simplicity of her means that overwhelms him with a sense of awe. This is what makes nature beautiful to him, Harping: the simplicity of the materials which make so many patterns, the unity under the surface chaos. Unity is the scientist's definition of beauty, and it makes nature beautiful to him all his life.

SIR EDWARD: But Potts, the sense of being at one with nature is not the prerogative of scientists. We all have it at times, and when we have it we find nature beautiful and inspiring.

POTTS: We all have it at times, Sir Edward, as a strangled, unformed and unfounded experience. But science is a base for it which constantly renews the experience and gives it a coherent meaning. I find the sunset beautiful even though I am colorblind—which is the shortcoming that I was trying to hide from you earlier this evening. But Harping, you see, does not find the sunset beautiful, and he does not want to talk about it, because he has no base for his incoherent sense of awe. Harping can only talk about human works of art.

*

HARPING: For the first time Potts has said something that impresses me. He has said that science gives him and his colleagues a sense of the unity of nature which is better founded than the shy moments of euphoria which you and I are swept by, Sir Edward. That makes sense to me; I give him that.

SIR EDWARD: I am glad to hear you say it, Harping.

HARPING: It makes sense to me because it underlines that Potts and I are doing different things. He is analyzing the workings of nature, and that would be a pitiful chore if nature did not move him and fill him with wonder. I, on the other hand, am fascinated by the expressions of the human spirit. That is a different fascination. Potts is quite right; nature alone, nature without the human spirit, means nothing to me, and I avoid discussing her if I can. I tried to avoid talking about the sunset this evening. It was acute of you to spot that, Potts, and I plead guilty.

POTTS: That's very handsome of you, Harping. I am proud to have convinced you.

HARPING: Not so fast, Potts, not so fast. You have convinced me that your students do not merely copy what you tell them; but that they unravel the links in nature for themselves, and that this experience makes them feel that nature is simple, profound and beautiful. Am I putting that fairly, Potts?

POTTS: Very fairly.

HARPING: But if that is so, Potts, where is there room for the individual judgment? You have a description of nature, your students argue about it, they have to see for themselves how intimately it hangs together; but at the end of the argument, they accept the description. The argument adds nothing—nothing for you and nothing for the student except to teach him the connections in nature. What was the point of the argument? There is nothing to argue about. This is how nature goes, take it or leave it. Nature is not open to argument and she cannot be enriched by the play of individual judgments and the human analysis of motives. Nature works and that is beautiful. Period. Can you understand now, Sir Edward, why I have no truck with such a definition of beauty?

SIR EDWARD: You must leave me out of this, Harping. I shall bore you, and anger you too, if I tell you that I prefer the sunset even to Shakespeare's *Sonnets*.

POTTS: But I don't, Sir Edward; I don't prefer the sunset; and I will not let Harping foist it on to me. You see, Harping talks as if there is only one vision of nature—

HARPING: I don't talk about a vision at all. I say that there is one nature, and you either describe her or you fail. There is no other choice for you.

POTTS: I will try to find a neutral word. Harping does not accept the word 'vision,' and I do not accept the word 'description' of nature. Let me try 'picture of nature.' Will you allow that, Harping?

HARPING: Let me hear what you have to say.

POTTS: I say that you talk as if there were only one picture of nature, a standard photograph which I and my students and my colleagues all have, now and forever, amen. But science isn't a bible—not even a bible of facts; and not even the facts stand still or look the same to two different students. Isaac Newton had a vision of nature before I did—I beg your pardon, a picture of nature. It was not the same as mine is. Tell me, Harping, was that a fault in Newton?

HARPING: Yes, it was. There is only one nature; you imply that when you are awestruck by her unity. And if there is only one nature, then there is only one correct description of her. Anything short of that description is an error, and even if the error is Newton's, it has no interest except as history.

POTTS: No, Harping. There is no correct description of nature. Nature is more subtle, more deeply intertwined and more strangely integrated than any of our pictures of her—than any of our errors, Harping. It is not merely that our pictures are not full enough; each of our pictures in the end turns out to be so basically mistaken that the marvel is that it worked at all. If Relativity is right, then the marvel is that Newton's mechanics worked at all.

HARPING: And what becomes of the unity of nature, Potts? Where are the connections? Where is the profound and underlying organization among these errors?

POTTS: But nature *is* the unity, Harping. We do not have to invent that. It exists out there, visibly, as its own proof. And it strikes us with awe exactly because it is so rich; because it can contain and embrace and be mirrored in all the pictures—Newton's as well as Einstein's, and Rutherford's atom as well as the wave equation of Dirac. Do you suppose that my students think that Rutherford was a fool because the atom has turned out to be more subtle than he pictured it fifty years ago?

*

HARPING: I don't understand you, Potts. You have a right to be sentimental about Rutherford yourself, of course; you probably remember him. But why bother your students with him? They want to know the facts; they want to know what the atom is like, truly. Why should they be interested in the past errors of legendary heroes? Rutherford is just a piece of history to them, a quaint allusion—another of your literary tags, like Samuel Butler, that is intended to make science look gentlemanly and cultured.

POTTS: Rutherford is part of the evolution of physics, of

that wonderful collective work, as the Elizabethans are part of the collective evolution of literature. And Rutherford was the greatest of his contemporaries, as Shakespeare was the greatest of the Elizabethans. To be ignorant of what Rutherford believed is as uncouth as not to have read Shakespeare. It would be uncouth in my students, and it's uncouth in you, Harping—and in you too, Sir Edward, if you claim any share in the culture of England in the twentieth century. Good God, how can you ask me to speak up for England at conferences, and quote Shakespeare, when you think that Rutherford is only a quaint figure in the history of science, like Dr. What's-his-name in *Tristram Shandy?*

SIR EDWARD: That is rather severe, Potts. However, I am here to note and consider criticism, so you must say what is in your mind. Is Potts's stricture fair, Harping?

HARPING: No, it is not fair, Sir Edward. It is a play with words, and no more—a skillful cliché, a cliché from the weeklies and the smart Sunday papers, but still a cliché. There is no ground of comparison between Rutherford and Shakespeare, and they do not belong to our culture in any similar sense. How can you draw an analogy between them, Potts, without blushing for your manipulation of clichés? Shakespeare did not use language in that way; he used it to express the essence of human motives, to evoke them and to command them, so that his language in itself created a judgment of the condition of man. Do you dare to ascribe such creations of the human spirit to any scientist, however powerful his mind? He may be a master in experiment, an eagle in observation, and subtle as a spider in spinning his theories; but he remains earthbound—bound to the sunset if you like, bound by the inhumanity of nature. Potts, how can you call any scientific mind, even Rutherford, a Shakespeare? That is nothing but cheap journalism.

SIR EDWARD: Don't answer him yet, Potts. I want to hear what you're going to say. So give me a moment, please, to taste the claret first. Yes, excellent, it will do very well with the veal. I apologize for the laminae of veal, gentlemen; but you know how the Swiss are about meat. A small country making money naturally likes conspicuous display, such as these acres of veal on our plates. And the Swiss are too greedy to let the display go to waste; they eat the meat, every crumb of it. If Thorstein Veblen had been Swiss and not Scandinavian, he would not have invented those grand theories about conspicuous waste.

*

POTTS: I should like to sustain my comparison, Sir Edward.

SIR EDWARD: You must, Potts, you must. But let me remind myself first how we got to it. Harping had granted you that scientists have cause to feel in their work a sense of unity with nature. But then, he asked, if nature is the criterion, how can any scientist claim a personal vision? How can you compare the free imagination of Shakespeare with the earth-bound inquiries of Rutherford? Go on, Potts, I've found my place.

POTTS: I should like to explain (and I had better do it in detail) why I think that a great scientist's picture of nature is as truly a creation and, yes, a vision, as is Shakespeare's vision of the human state. But I will grant Harping one thing: that Shakespeare is not a good subject for the comparison he has challenged me to make. Shakespeare is not a good subject, not because he is Shakespeare, but because he is a writer. I would like to make the comparison with some other kind of artist; a sculptor, perhaps, or a composer, or a painter —someone who does not use words; someone who can be appreciated in every language. Will you give me a painter, Sir Edward?

SIR EDWARD: Leonardo da Vinci.

HARPING: I won't agree to Leonardo; he was too close to

science to serve as a fair comparison. If we are to have a painter, he must be a pure painter.

POTTS: Choose a pure painter, Harping.

HARPING: I will choose Rembrandt.

POTTS: Very well, I will compare Rutherford with Rembrandt. They make a very good pair. They were both dedicated men, both absorbed in their own activity to the point of passion. You recognize them at once, these creative men, by their patience, their preoccupation, their physical immersion inside the things they make. Their work is a world for them.

SIR EDWARD: You must not beg the question, Potts. You have slipped in the word 'creative'. I have no doubt that Rutherford was always busy with his work, and single-minded like Rembrandt. I accept your *bon mot*: their work was a world for them. Men like that are fortunate; they are happy in their work, even in adversity—absorbed, happy and fulfilled. I accept all those words, Potts. But creative? Have you given me any ground for the word 'creative'?

POTTS: Rembrandt and Rutherford both made something, Sir Edward. And what they made was original. Isn't that the definition of creation?

SIR EDWARD: Now I'm not happy with your word 'original'. It's too narrow for what we are trying to grasp. No doubt Rutherford had an original mind; but the mind is only a small part of a rounded personality. What I want you to tell me is quite downright: Is a great scientist creative, not merely as an intellect, but as a man? And if you shift your ground to the word 'original', you do not satisfy me; you merely make me use your new word to ask the old question. Is the physics of Rutherford original in the same deep sense that the paintings of Rembrandt are original? Does the physics of Rutherford express the whole man, head and heart, in the same deep sense that the paintings of Rembrandt evidently look at us out of his whole self?

POTTS: I think it does, Sir Edward. Rutherford's physics is as subtly woven together, as delicately compounded of both sides of him, fact and imagination, as every human activity is. Neither his science nor any science is all head, all fact; and Rembrandt's painting is not all heart either, as if it were a rainbow cloud of fancies without a base of fact.

HARPING: Clichés, my dear Potts, clichés. Sunday journalism.

POTTS: Let me finish, Harping. I was talking about fact and imagination, in physics and in painting. You will agree that Rembrandt as a painter was wedded to the facts. In one sense, his paintings are an exact description of what he saw. Rembrandt's paintings are not photographs, certainly; but they are representations, and they were intended by Rembrandt (and accepted by those who commissioned them) to represent reality. In this sense, Rembrandt's paintings are every inch as factual as Rutherford's descriptions of his experiments.

SIR EDWARD: Go on.

POTTS: But of course, Rutherford's reputation was not made by his description of the experiments. It was made, like Rembrandt's, by his interpretation: his interpretation of what lay hidden below the surface reality and which the experiment or the painting revealed. One experiment, one painting, pointed the way to the next, until they wove together a network of interpretations which made a single image.

SIR EDWARD: An image of what?

POTTS: In Rembrandt's self-portraits, an image of himself. In Rutherford's atomic experiments, the extraordinary and unbelievable image of the atom as a minute solar system.

SIR EDWARD: And these images are not true?

POTTS: What is truth? Is the sequence of Rembrandt's self-portraits the truth about himself? The atom is not truly a small solar system. Rembrandt and Rutherford both began with the world of fact. Both of them put down in the first place what they saw; and both of them transformed what they saw into something more delicate and more illuminating than any mere record of the facts. The painter's portrait and the physicist's explanation are both rooted in reality, but they have been changed by the painter or the physicist into something more subtly imagined than the photographic appearance of things. The portrait and the theory are, both of them, at once more original and more personal than the commonplace eye of the camera.

SIR EDWARD: You are now equating the words 'original' and 'personal'.

POTTS: Of course—because to see something in an original way, we must see it in a personal way. This is obvious about paintings. Every intelligent layman, when he looks at a painting, knows that it combines two elements. He knows that it is a picture of reality, and he also knows that it is reality as one man sees it. The painting is two things: a picture and a vision. But the layman has still to learn, you both have still to learn, that what is true of painting is also true of physics—that physics also, I have to repeat it, is constructed by men, not machines. Of course, physics is an account of physical reality. But the turning points in this account, the moments of great discovery in physics, are flashes of vision when a single man sees a new link between different and apparently unrelated aspects of reality. In this visionary moment, the great scientist lays bare a new linking, a new pattern among things. And his vision is as imaginative, as much a creation, as the painter's vision.

*

SIR EDWARD: Then why does the scientist have no difficulty in sharing his vision with others? If Rutherford saw with so personal an eye, why were his findings accepted by other scientists as soon as they were published? I seem to remember that Rutherford went to the Lords; but Rembrandt died in poverty.

POTTS: There have been painters who went to the House of Lords, too, I think; and great scientists who died misunderstood and neglected—Mendel, for one. But you are right: the scientist can make his vision acceptable in a more systematic way than the painter can. That is, there exists a method of handing on the findings of physics to other people so that they usually accept them; and there exists no such assured method in painting. But the method is only an exposition; it has nothing to do with the way that the finding entered the mind of the discoverer. The finding, in physics as much as in painting, remains a personal illumination; we help every student to re-create it, but we cannot teach one student to create it. Rutherford's model of the atom was not a fact simply concealed in nature and waiting for any Tom, Dick or Harry to fish it up. Rutherford's absurd and wonderful model of the atom was an imaginative discovery, a highly personal way of seeing nature—even though Rutherford was then able to persuade a thousand other physicists to see nature in his way.

SIR EDWARD: You are very persuasive, Potts; and of course, you are convinced yourself. Yet you still perplex me. I still don't know how Rutherford's way of looking at things was different from that of other physicists who were, I suppose, doing similar experiments at the same time. Obviously, Rutherford was different; but was he different as a person? Was he not simply cleverer?

POTTS: In his critical experiment, Rutherford and his students fired helium particles at a thin layer of atoms. He was astonished to find that some of the helium particles did not go through, but came back at an angle. Rutherford expressed his surprise by saying that he felt as if he had fired an artillery shell at a sheet of tissue paper, and it had bounced off.

He concluded that the tissue paper must be full of holes and lumps—and that each atom must be full of holes and lumps. That is how Rutherford came to the idea, from this metaphor, that the atom is not an electric cloud. Each atom has a structure; there is an inner lump or nucleus, and then around it a set of light electrons. The electrons circle the nucleus at quite large distances, just as the planets circle the sun a long way off. And whether things go straight through the atom, or whether they bounce back at strange angles, depends simply on whether they brush through the open space, or whether they actually hit the heavy nucleus.

SIR EDWARD: I understand the picture, and I see how it derives from the metaphor. Indeed, it seems to me that there were two metaphors in Rutherford's mind: first the artillery shell, and then the solar system. You will say, Potts, that the metaphors were individual to him, and that linking them together was most individual.

POTTS: To anyone who knew the bluff colonial manner of Rutherford, the roughness and the twinkle, the solemn sense of pulling his own leg, everything in his discovery of the structure of the atom is of a piece. The metaphors are as much a part of Rutherford's personality as the idea of the experiment. They all have the thumbprint of a man to whom the core of things had a strong and clear outline—even if it was as minute as the nucleus; and who judged things by tests which had to be equally clear and strong. Everything here is as individual and as human as Rembrandt. Do you remember the pictures of his mistress that Rembrandt painted as an old man, and that no one would buy? They have the same obstinate tangle of feelings, rough and tender at the same time, that Rutherford had for his mistress, nature.

SIR EDWARD: In short, Professor Potts?

POTTS: In short, a rich, broad body of science is like a rich, broad body of painting—and of poetry, too, though I have shirked the discussion of that. It springs from the vision of a rich, broad person, head and heart together, and the surface facts deepened and transfigured by the creative metaphors of the imagination. In short, Sir Edward, if I have to give you a single answer to the question, 'What makes a great scientist?' then I say that he is made as every creative man is made, painter or novelist, musician or poet. A man can only create something of which he has a vision. Rutherford was as much a visionary as was Rembrandt, or Beethoven, or Shakespeare. And that has already been said by a poet. William Blake said, as if he were speaking about science: 'What is now proved was once only imagin'd.'

HARPING: Of course; the tag from Blake. No lyrical account of science is now complete without a quotation from William Blake. And when Rutherford's vision of the atom turns out to be a hallucination, when the picture is proved wrong, does Blake have a wise saw for that, too?

POTTS: Yes, he does, Harping. Blake knew that human knowledge goes forward by an endless march of false steps.

SIR EDWARD: I must stop you, Potts. I ordered a dish that will not wait, and I see it approaching. A cheese soufflé: this is something that the Swiss do very well. I will give you the first helping, Potts, and you are to make no sound for a while but the sound of eating. You have done very well. You have made me see into a great discovery; you have made me catch the undertones in it, and now I know what you mean when you call it a work of creation. He has been very enlightening, Harping, don't you agree?

*

HARPING: Yes, Sir Edward, I agree. He has put his case well, and he has made me weigh things that I held too lightly before. I shall not forget what you have said, Potts; I shall think about it.

SIR EDWARD: But, my dear Harping, but—do I hear a 'but'

in your voice? What is your 'but'? You have almost got as far
as saying 'yes' to Potts; there must be a 'but' to hold it down.

HARPING: Of course there is, Sir Edward; a large 'but,'
a 'but' as large as life. My 'but' is, haven't we lost sight of life?
It is very fine to praise the vision of Rutherford and Einstein, and to trace the personality in their discoveries. But
how does the discovery enter, not their lives, but ours? How
does the vision express itself in the realities of the human
condition now?

SIR EDWARD: Yes, Dr. Harping? How?

HARPING: In the hydrogen bomb, Sir Edward. That is how
Einstein's insight and Rutherford's metaphor enter the life
of modern man: not as a Rembrandt, but as a bomb. The
atom may have symbolized the unity of nature to them; to us,
the very word is a threat to tear nature apart. Science may
have been a dream of life to them; to us, who breathe the
reality, it is the threat of death.

SIR EDWARD: Oh come, Harping, that's rather hard. We
have enough of the hydrogen bomb at the conference. Must
you drag it in here? Surely our talk this evening is outside
politics; I'd rather not turn it into a Trafalgar Square sitdown. I don't think the hydrogen bomb is in place here.

HARPING: But it is in place, Sir Edward; it's squarely in
place here. You call the big bombs a piece of politics, and
that reassures you. You can then shut your mind to them,
and write off as cranks those who won't shut their minds.
You've missed the point, Sir Edward: the point of the pickets
and the sitdown squatters and the bearded cranks. The hydrogen bomb is not a piece of politics to them. It is the monstrous climax of an invasion that has mounted and mounted
since the day they were born.

SIR EDWARD: I don't follow you, Harping.

HARPING: People are protesting against the invasion of
their lives, not by this death-dealing machine or that, but by
the machine itself—by the machine as a form of spiritual

death. Do you really think in Whitehall, Sir Edward, that all those sandaled men and homespun women are marching from Aldermaston because they are afraid to die in a sizzling sunset flash? No; they are protesting against the mechanics of all that is happening to life: against the domination of life by mechanics. They are not afraid that mankind will perish; they are afraid that *humanity* has perished already under the big wheel of scientific progress. Potts, I will give you a quotation from Blake, too: take it to heart, and you, Sir Edward. 'A Machine is not a Man nor a Work of Art; it is destructive of Humanity & of Art; the word Machination.'

SIR EDWARD: Very well, Harping, I will withdraw my remark about politics. I accept that you at least (I will not include the marchers)—I accept that you are objecting to the mechanization of peace as well as of war. Yet, I still ask, is that relevant to our discussion this evening? The sun is down and the stars are up; the atomic processes of nature flash and glitter in the sky, a fusion and profusion, in honor of Einstein and Rutherford. Is it right to blame them for the technical uses and abuses to which their discoveries have been put by others? After all, we didn't ask Potts to defend technical civilization. We asked him to give us an exposition of science as an imaginative culture.

HARPING: Whom else shall we blame? I am sure Einstein was a good man, Sir Edward, just as I'm sure Christ was a good man. But if you lived under the Inquisition, you would not be consoled by the thought that Christ did not plan it. I live under the shadow of destruction: physical destruction and spiritual destruction by the material domination of science. And I am not consoled to know that Einstein did not plan that. You spoke just now of science as a culture; Potts is fond of speaking that way. A culture is made up, intimately made up, of two related parts: an activity, and a vision—a way of doing things, and a way of thinking and feeling about them. I do not like to stand by and see our scientific civilization *do* things inhumanly, simply because Potts tells me that great scientists have a vision.

POTTS: And what would you like to do, Harping?

HARPING: I would like to destroy the idol of technical advance. I would like to get rid of the preoccupation with productivity, with material standards, with hygiene and technology and progress. I would like human beings to stop worshiping the machine. Do you know what all this talk about a scientific culture does to the arts? It makes an architect famous because he says that a house should be a machine for living. I do not want to give my mind to living; I want to give it to life. And I do not want a house to be a machine; I want it to be a home.

POTTS: Even if home is a slum, Harping? Even if hygiene and progress have to get the bedbugs out of the woodwork? You talk as if material standards were something vulgar that the poor ought to be ashamed to enjoy, like sex and cheese soufflé. Do you really want me to be ashamed of the help that science has given in feeding people and housing them and giving them transport and print and sanitation? Do you think people would be better if they were less healthy?

HARPING: They would be better if they thought less about their health. And they would certainly be better if they thought less about their wealth.

POTTS: You forget what thinking has done for people: thinking about their health, and their station, and their ambitions and gifts. This is the drive behind the material advance, this is why people struggle for the comforts that you affect to despise: because they want to lead fuller lives, to become individuals—not just because they want to *have* more, but because they want to *be* more. What was your father's trade, Harping?

HARPING: Why should you want to know? He was a tailor. What was yours?

POTTS: Mine was a groom. He really was an Irish groom.

Do you think that the sons of the tailor and the groom would have gone to Cambridge a hundred years ago? Or that even now in Mexico, you would be a critic and I a biologist? Don't belittle our industrial civilization, Harping. It has many aberrations, all the way from bombs to stomach ulcers. But it does one thing better than any other civilization has done. It gives young men the chance to make the most of their talents.

*

HARPING: You are always thinking about people's talents. I am thinking about their lives—their personal relations, their loves and rejections, their community with themselves and with others. These are the values that industrial society destroys. And in their place it puts nothing: a blank, a mass entertainment which demands neither their attention nor their participation, a television screen which grimaces so inanely that it would be better blank. Whitewash your brain blanker than blank. My father was not prosperous enough to go to public entertainments, and would not have had the leisure if he had been. He didn't see cowboy films or play bingo, he didn't have a betting shop around the corner and he didn't fill in the pools. And if these constitute the total benefit of prosperity and leisure, then I will say, as Dr. Leavis says: the felicity it represents cannot be regarded by a fully human mind as a matter for happy contemplation.

POTTS: My dear Harping, rubbish! How can you be so ignorant of the lives of your contemporaries? As if working men did nothing but drink and bet and wait for you to tell them what to do with their leisure. And how solicitous you are for their leisure! Is yours so well spent? Is yours, Sir Edward? No one ever seems to have any trouble with his *own* leisure. It is only other people's leisure that is a social problem. The fact is, Amos Harping, you are a puritan. You cannot bear to have other people enjoy themselves in any way but yours. That is what you find distasteful, that is what outrages you about the success of technology. It makes it too easy for people to be well off, well fed, just well. Your soul is still in the age of famine; you have not come to terms with the prospect of plenty. You believe that the values of life come from denial, not acceptance. And when you see other people idle, how mealy-mouthed you become—like a bishop, or like Sir Edward St. Ablish writing to *The Times* about the welfare state! Did you really hear your quotation? 'The felicity it represents cannot be regarded by a fully human mind as a matter for happy contemplation.' What is a fully human mind, Harping?

HARPING: It is a mind that does not lull itself with your ready-made optimism, like a liberal weekly. It does not take the culture of the Sunday papers, even the highbrow Sunday papers, to represent the best that is thought and known in our time.

POTTS: Do stop these evasions, Harping. And stop nagging at the Sunday papers. They do a simple job: they take the place of the Sunday sermon and the Bible reading, and they do it well. And their style is better than yours. I asked you a question in your own words: What is a fully human mind? Give me a positive answer, not a negative one.

HARPING: It is a mind which is occupied with the human state, as an individual in a living community. It sees its life and its work in relation to those of its neighbors, as an exchange among equals. It is not concerned to keep up with the Joneses in jobs or cars or washing machines. That may seem a primitive life to you, Potts. Yet, if I have learned anything from my students, anything from my academic fellows and, yes, from my master, it is that this primitive life is a good life. Potts, Sir Edward—I will ask you both the question that Dr. Leavis asks: Who will assert that the average member of a modern society is more fully human, or more alive, than a Bushman, an Indian peasant, or a member of one of those poignantly surviving primitive peoples with their marvelous art and skills and vital intelligence?

POTTS: Who will assert *what?* I assert it, Amos Harping. I assert that the average man who drove our train up here is more human and more alive than any of your poignant primitive people. The skills of the Bushman, the vital intelligence of the Indian peasant? You are tipsy with sentiment, Harping, or you would not compare them with the man who reads your proofs. The Bushman and the Indian peasant have not been cowed by science, Harping. They have failed in culture: in making a picture of the universe rich enough, subtle enough—one that they can work with and live by beyond the level of the Stone Age. They have failed because they did not create a mature view of nature, and of man too, Harping. My God, you talk, you dare to talk, of their marvelous art. Since when have you been an admirer of Bushman art, Harping?

HARPING: That's a pointless question, Potts. I have always admired it.

*

POTTS: Then why did you give me Rembrandt when I asked you for a painter? Why do you, Dr. Amos Harping, lecture to your students about George Eliot and not about Indian folk poetry? Because you know that Rembrandt is a more mature artist than any Bushman, and George Eliot than any folk poet. I don't understand you, Harping. How can you be so blind to the evidence of your own practice? You try to enrich the emotional appreciation of your students—how? By discussing Shakespeare with them; and Joseph Conrad, and D. H. Lawrence. How does it happen that Shakespeare was not born in the bush?—or Conrad or

Lawrence? Every work that you present to your students as masterly, as profound and sensitive, was produced in a society with a high standard of technical sophistication.

SIR EDWARD: Say that again, Potts. That is a new thought to me, and an interesting one.

POTTS: Yet surely it should be a very immediate thought, Sir Edward. Here is Amos Harping preaching to us that the greatest works of man express his concern with his own life and condition, with his own humanity. And that concern and that humanity, he tells us, are eroded by the advancing tide of technical civilization. Indeed, in a heady moment he says that modern man is less fully human, on the average, than the poignantly surviving primitive peoples. Very moving! Hitler used to say it too, and called it 'Blood and Soil'. But do the great works of man come from the poignant primitive peoples? Do they even come from the poor whites of Tennessee, from the stony fields of Spain, or from the starveling fisheries of Sardinia? Of course not. Dr. Harping lectures on none of these. Then where were the searching analyses of the human condition made? Where were the books written that most deeply express and explore the humanity of man? In the Athens of Sophocles, in the Florence of Dante, in the England of Shakespeare. Yet these were not simple, ascetic societies; on the contrary, they were the most highly developed technical and industrial societies in history. That is where the human mind realized itself most fully: in the cities that stand at the peak of technical achievement in their time. That is where great men flower, Harping, whatever you may say about the average man. The evidence is there, in the books you recommend to your students for their human insight.

HARPING: The evidence is a trick, Potts. You read it falsely. It is a mere trick to say that Athens and Renaissance Italy and Elizabethan England were technically advanced. They were not dedicated to the worship of gadgets. They were not slaves to the technological search for hygiene and personal comfort.

POTTS: Archimedes was having a bath when the idea of specific gravity came to him. That was in Syracuse, a colony of Greece. If Archimedes had been born in the culture of Tibet instead of Greece, he would no doubt have been just as shining and incisive a mind. But he would never have jumped out of his bath and shouted 'Eureka!' because he would not have had a bath.

SIR EDWARD: Is that very subtle, Potts? Surely Archimedes in Tibet would have failed to make his discovery for deeper reasons. He would have been thwarted by the indifference and lack of understanding, by the emotional hostility of his surrounding civilization.

POTTS: Of course, Sir Edward: he would have been thwarted by Dr. Harping. And Harping is against baths. You have just heard him: they represent slavery to the technological search for hygiene and personal comfort. The emotional hostility to science is all of a piece. Harping's puritanism is the same here and in Tibet; and it hates the material bath and the vision of the atom, both together.

*

HARPING: I refuse to discuss baths in Tibet, about which none of us knows anything. And I refuse to be drawn away from my point. Potts's reasoning is simply false. Yes, greater works were produced in Greece and her colonies and in Renaissance Italy and in Elizabethan England than elsewhere. But these places, these islands in history, were not technical societies in our sense. They were not industrial treadmills as our society is. By your evidence, Potts, our society ought to be producing its Sophocles and Dante and Shakespeare.

Where are they? Or are you going to claim D. H. Lawrence, because he was born in industrial Nottingham—and hated it? All that our scientific, hygienic, technological society did for Lawrence was to hound him to death.

POTTS: You talk as if Lawrence had been persecuted by a conspiracy of physicists. The people who persecuted Lawrence were the same people who abused Einstein: not physicists but Philistines. Lawrence was driven out by the Establishment, by people like Sir Edward here—let's face it. And people like Sir Edward are just as hostile to new science as they are to new art. As a matter of fact, Philistines like Sir Edward are hostile to any kind of science; and you ought to be ashamed to join with them, Harping.

HARPING: I am not on the side of the Establishment. Good God, I have fought Sir Edward and the Philistines all my life.

POTTS: All the same, Harping, the ranks of the Philistines are full of literary critics, and they are not full of scientists. Perhaps that is the strength of science, and its inspiration to the young: that those who are dedicated to it are themselves creators of science—modest creators, but creators. There are no critics in science, no high priests who only expound and guard the godhead. I am embarrassed by your form of priesthood, Harping. The gods you guard are always dead: even D. H. Lawrence. I wish I had heard you once praise a living writer. Don't you have any living poets or novelists at your university?

HARPING: That is a fashionable heresy: that a literary teacher ought to qualify as a poet, or a commercial novelist. I don't share it.

SIR EDWARD: Then you must not be surprised, Dr. Harping, if for once we Philistines side with Potts. I think that you would have liked to be a poet, but you have not enough joy and zest and confidence. You are a self-defeated man, Harping: the kind whom John Dryden meant when he said, 'The corruption of a poet is the generation of a critic.'

POTTS: Don't abuse Harping, Sir Edward. It was I who rounded on you, not he.

SIR EDWARD: I know that, Professor Potts, but I haven't got to the bottom of you yet. I don't think I shall. But there is still the issue that I want to see settled: something that you said that struck me as quite fresh. You said that the great works of art of the past were always created by people with high technical standards of living. Harping says that they were not.

POTTS: Harping is wrong. The great ages of art in the West were in Greece, and Italy in the Renaissance, and the England of Elizabeth and the Restoration. Those were also the places that were technically most advanced, most inventive, most excited by the adventure of progress. And people lived better there, ate better, had better tools than anywhere else in the world as it was then.

HARPING: Yet I say that these were not technical societies, machine societies, in the modern sense.

POTTS: You are wrong, Harping. You sentimentalize the past—the Elizabethan past as much as the Bushman past. You think that a Renaissance water wheel was charming and a dynamo is not, just because the water wheel was made of wood. And you really think that the silver mines of Greece were more human, more cozy, than a coal mine today. You are wrong. Those people made a high standard of living for themselves exactly as we do, by working for it, working hard and with the best tools that man had then invented.

HARPING: They didn't play bingo and read comic strips.

POTTS: Didn't they, Harping? How do you know? The evidence is that they did whatever gave them pleasure, and that everything in life gave them pleasure. Are the mystic numbers of Pythagoras different from bingo? Are the dirty pictures on the Greek vases different from ours? Those people saw life as one, the whole of intellectual life—numbers and pictures, the lever and *The Iliad,* the shapes of atoms and the great plays and the Socratic dialogues. You are the one who divides life into little pieces, Harping, each piece as big as the bite of one critic. But the creative drive, the lust to make and discover, the itch of originality—*that* gets a whole age by the scruff of the neck, and then art and science and technology go forward headlong, higgledy-piggledy, as confused as Leonardo da Vinci and as universal as Aristotle. You know very well, Harping, that Marlowe didn't spend his time with literary critics; he spent it with mariners and scientists and adventurers. Galileo's father was a musician, but he did not rail at his son for taking to science. In those ages, the great ages, technical wellbeing and art and science were all one; and so they can be today, if we let them. We have creative men in art as well, I have no doubt, if we let them breathe— if we open their minds to our world, not the world of the Bushman. I shall never change you, Sir Edward; you *are* the past.

SIR EDWARD: Oh what nonsense!

POTTS: But Harping has no business to play the obscurantist. When Galileo wrote his *Dialogue on the Great World Systems,* the Church broke him: that was you, Sir Edward, your Establishment—they called it the Holy Office in 1633. What side were you on, Dr. Amos Harping, when the Philistines made that heretic recant?

*

SIR EDWARD: Gentlemen, gentlemen, I have quite forgotten myself. I ordered wild strawberries as a *bonne bouche,* and the last drops of the Pouilly Fumé to spill over them. And coffee, and a glass of brandy. How an evening slips by! We have thirty minutes before the last train. It has been a

wonderful evening—I beg your pardon, Harping: a splendid
evening. Thank you both. Let us give the last of it to the
stars, and the strawberries, and the brandy. All but one
minute, Professor Potts. I cannot go back to the conference
without asking a one-minute question. Potts, I still don't
know what a molecular biologist is, and now I never shall.
Don't tell me. But tell me this: why are you scientists so successful? Why are you conquering the world, in spite of me?
Why are you conquering the young, in spite of Dr. Harping?

POTTS: Because ours is a vision and an activity together,
Sir Edward. That is how Harping defined a complete culture, and it is. science is a culture. We are the visionaries of
action; we are inspired with change. We think the past pre-
serves itself in the future of itself, the way Isaac Newton is
changed and still preserved in Albert Einstein. We are the
culture of living change.

SIR EDWARD: And is that enough, Professor Potts? I belong
to the Establishment, which means, you say, that all my life
I have believed that tradition is enough. Well, suppose I
am wrong. But does it follow that change is enough? Any
change?

POTTS: No, Sir Edward, not any change. What we are
changing is the division of life. We are making a unity—a
complete culture, a unity out of variety. I will tell you what
a molecular biologist is. He is a man who unravels the
secrets of life by using the tools of physics. He shows—we
have shown—that the structures of biology become in-
telligible when we treat them, not as strings of mysteries, but
as strings of molecules. Those are the changes in the picture
of the world that we drive for. That is the universal unity in
which we believe. One of my fellow scientists wrote a poem
once about art and science and the unity of all things. May I
recite it, Sir Edward?

SIR EDWARD: You ought to ask Dr. Harping. But I will
listen anyway.

POTTS:
I, having built a house, reject
The feud of eye and intellect,
And find in my experience proof
One pleasure runs from root to roof,
One thrust along a streamline arches
The sudden star, the budding larches.

The force that makes the winter grow
Its feathered hexagons of snow,
And drives the bee to match at home
Their calculated honeycomb,
Is abacus and rose combined.
An icy sweetness fills my mind,

A sense that under thing and wing
Lies, taut yet living, coiled, the spring.